STUDY GUIDE

for the First Edition of

PUMPING APPARATUS DRIVER/OPERATOR HANDBOOK

John Joerschke
CONTRACT WRITER

Cindy Pickering
PROJECT MANAGER/EDITOR

Published by
Fire Protection Publications Oklahoma State University
Stillwater, Oklahoma

The International Fire Service Training Association

The International Fire Service Training Association (IFSTA) was established in 1934 as a "nonprofit educational association of fire fighting personnel who are dedicated to upgrading fire fighting techniques and safety through training." To carry out the mission of IFSTA, Fire Protection Publications was established as an entity of Oklahoma State University. Fire Protection Publications' primary function is to publish and disseminate training texts as proposed and validated by IFSTA. As a secondary function, Fire Protection Publications researches, acquires, produces, and markets high-quality learning and teaching aids as consistent with IFSTA's mission.

NOTICE: The questions in this study guide are taken from the information presented in the first edition of ***Pumping Apparatus Driver/Operator Handbook,*** an IFSTA-validated manual. The questions are *not validated test questions and are not intended to be duplicated or used for certification or promotional examinations;* this guide is intended to be used as a tool for studying the information presented in ***Pumping Apparatus Driver/Operator Handbook.***

ISBN 0-87939-169-3

First Edition

Printed in the United States of America 4 5 6 7 8 9 10

If you need additional information concerning our organization or assistance with manual orders, contact:
Customer Service, Fire Protection Publications, Oklahoma State University
930 N. Willis, Stillwater, OK 74078-8045
1-800-654-4055 FAX: 405-744-8204

For assistance with training materials, to recommend material for inclusion in an IFSTA manual, or to ask questions on manual content, contact:
Editorial Department, Fire Protection Publications, Oklahoma State University
930 N. Willis, Stillwater, OK 74078-8045
405-744-5723 FAX: 405-707-4112 E-mail: editors@osufpp.org

Table of Contents

Preface

This study guide is designed to help the reader understand and remember the material presented in **Pumping Apparatus Driver/Operator Handbook**, first edition. It identifies important information and concepts from each chapter and provides questions to help the reader study and retain this information. In addition, the study guide serves as an excellent resource for individuals preparing for certification or promotional examinations.

Much time and effort go into the design, development, layout, and printing of any publication. This **Study Guide for the First Edition of Pumping Apparatus Driver/Operator Handbook** is no exception. I would like to give special recognition to the following members of the Fire Protection Publications staff whose contributions made possible the technical accuracy and visual appeal of this publication.

John Joerschke, Contract Writer
Barbara Adams, Associate Editor
Susan S. Walker, Instructional Development Coordinator
Mike Wieder, Senior Publications Editor
Desa Porter, Senior Graphic Designer
Don Davis, Publications Production Coordinator

Cindy Pickering
Cindy Pickering
Curriculum Specialist

How to Use this Book

This study guide is developed to be used in conjunction with and as a supplement to the first edition of the IFSTA manual, **Pumping Apparatus Driver/Operator Handbook.** The questions in this guide are designed to help you remember information and to make you think—they are *not* intended to trick or mislead you. To derive the maximum learning experience from these materials, use the following procedure:

Step 1: Read one chapter at a time in the **Pumping Apparatus Driver/Operator Handbook** manual. After reading the chapter, underline or highlight important terms, topics, and subject matter in that chapter.

Step 2: Open the study guide to the corresponding chapter. Answer all of the questions in the study guide for that chapter. You may have to refer to a dictionary or the glossary in the **Fire Service Orientation and Terminology (O&T)** manual for terms that appear in context but are not defined. After you have defined terms and answered all questions possible, check your answers with those in the answer section at the end of the study guide.

✓**Note:** *Do not* answer each question and then immediately check the answer for the correct response.

If you find that you have answered any question incorrectly, find the explanation of the answer in the **Pumping Apparatus Driver/Operator Handbook** manual. The number in parentheses after each answer in the answer section identifies the page on which the answer or term can be found. Correct any incorrect answers, and review material that was answered incorrectly.

Step 3: Go to the next chapter of the manual and repeat Steps 1 and 2.

Chapter 1

True or False

5. True Federal regulations in the United States and Canada exempt driver/operators from having to obtain a CDL for driving a large truck pg 3

1 Who usually determines the required level of skills for a driver/operators?

a. National Fire Protection Service
b. State Law
c. Local jurisdiction
d. Federal ~~guidelines~~ regulations

2 What NFPA standard sets minimum qualifications for driver/operators?

a. 1001
b. 1002
c. 1201
d. 2001

6. Which of the following statements about driving regulations is true?

a. ignorance of driving laws and regulations is an admissable defense
b. most laws concerning motor vehicle operation are maintained at the state or provencial level
c. emergency vehicles are generally exempt from all speed limits
d. In most jurisdictions emergency vehicles are exempt from laws requiring vehicles

to stop for school buses ~~where~~ ~~flashing lights~~ whose flashing red lights indicate children are being loaded or unloaded

Chapter 1 The Driver/Operator

True/False

A. Write *True* or *False* before each of the following statements. Correct those statements that are false.

_______ 1. The NFPA *Standard for Fire Apparatus Driver / Operator Professional Qualifications* requires driver/operators who will be responsible for operating a fire pump to also meet the requiements of NFPA 1001.

____

____

_______ 2. Reading and writing are not important skills for driver/operators.

____

____

_______ 3. Driver/operators should be able to determine square roots and solve simple equations.

____

____

_______ 4. If a driver/operator's negligence leads to an accident, both the driver/operator and the fire department may be held responsible.

____

____

_______ 5. Federal regulations in the United States and Canada exempt driver/operators from having to obtain a CDL for driving a large truck.

____

____

Multiple Choice

B. Write the letter of the best answer on the blank before each statement.

____ 1. Who usually determines the required level of skills for driver/operators?
   a. National Fire Protection Association
   b. State Law
   c. Local jurisdiction
   d. Federal regulations

____ 2. What NFPA standard sets minimum qualifications for driver/operators?
   a. 1001
   b. 1002
   c. 1201
   d. 2001

____ 3. Which of the following statements regarding driver/operators is true?
   a. The physical demands of driving and operating an apparatus are light.
   b. NFPA 1582 requires driver/operators to have a corrected far visual acuity of 20/20.
   c. NFPA 1582 recommends rejecting the firefighter candidate who has a hearing loss greater than 25 decibels in any one of three frequencies.
   d. The driver/operator must be subjected to a periodic medical evaluation in accordance with NFPA 1500.

____ 4. Which of the following skills is ***not*** required for driver/operators?
   a. Reading
   b. Writing
   c. Math
   d. Welding

____ 5. Which of the following statements about the selection of driver/operators is true?
   a. In volunteer departments, the chief officer's judgment usually determines potential driver/operator candidates.
   b. A professional firefighter's promotion to driver/operator should be based solely on seniority or position.
   c. Individuals with truck-driving experience who are not firefighters can move laterally into volunteer fire departments as driver/operators with no additional training.
   d. NFPA 1504 outlines requirements for establishing a driver/operator training program.

____ 6. Which of the following statements about driving regulations is true?
   a. Ignorance of driving laws and regulations is an admissable defense.
   b. Most laws concerning motor vehicle operation are maintained at the state or provincial level.
   c. Emergency vehicles are generally exempt from all speed limits.
   d. In most jurisdictions, emergency vehicles are exempt from laws requiring vehicles to stop for school buses whose flashing red lights indicate children are being loaded or unloaded.

Identify

C. Identify the following abbreviations associated with the driver/operator. Write the correct interpretation before each.

________________ 1. NFPA

________________ 2. MDT

________________ 3. SOP

________________ 4. DOT

________________ 5. TC

________________ 6. CDL

Chapter 2 Types of Fire Apparatus Equipped with a Fire Pump

Matching

A. **Match to their definitions terms associated with types of fire apparatus equipped with a fire pump. Write the appropriate letters on the blanks.**

____ 1. Scaled-down version of standard fire department pumper

____ 2. Used to transport water to areas where water supply is inadequate

____ 3. Mobile water supply apparatus that act as a reservoir for some fires

____ 4. Operation where mobile water supply apparatus dump their loads into a portable water tank

____ 5. Apparatus used to control ground cover fires

____ 6. Apparatus that can be driven and discharge water at the same time

____ 7. Separate motor used to operate the fire pump

____ 8. Protects the front and undercarriage of an apparatus and extinguishes small fires in short vegetation as the apparatus is advanced

____ 9. Separates water tanks into compartments

____ 10. Step-up transformer that converts the vehicle's 12- or 24-volt DC current to 110- or 220-volt AC current.

____ 11. Allows multiple hydraulic tools to be connected to the end of a supply hose

____ 12. Method of categorizing pumpers by their capabilities

a. Ground sweep nozzle
b. Manifold block
c. Generator
d. Auxiliary engine
e. Nurse tender (tanker)
f. Booster apparatus
g. Tender
h. Apparatus typing
i. Pump and roll
j. Inverter
k. Water shuttle
l. Baffle
m. Initial attack apparatus

True/False

B. Write *True* or *False* before each of the following statements. Correct those statements that are false.

_______ 1. The main purpose of a fire department pumper is to provide water at an adequate pressure for fire streams.

____

____

_______ 2. Industrial fire pumpers frequently have pump capacities in excess of 2,000 gpm *(8 000 L/min)*.

____

____

_______ 3. Most large-scale industrial foam pumpers use some form of around-the-pump foam proportioning system.

____

____

_______ 4. Elevating water devices cannot be used to apply fire streams to the lower floors of a building.

____

____

_______ 5. Midipumpers usually are built on a chassis of over 12,000 pounds (5 443 kg) GVW.

____

____

_______ 6. One approved method for making a moving fire attack is to have firefighters ride on the outside of the apparatus and discharge water as the vehicle is driven.

____

____

_____ 7. Individual master stream turrets that discharge 26,000 gpm *(98 420 L/min)* are common on fire boats.

_____

_____

_____ 8. One advantage of equipping an aerial apparatus with a fire pump is that the apparatus can be used to protect itself in high radiant-heat situations.

_____

_____

_____ 9. It is common for the fire pump and water tank on rescue apparatus to be inside one of the compartments.

_____

_____

_____ 10. Inverters on pumpers are most commonly used to provide power for portable equipment.

_____

_____

_____ 11. Generators are the most common power source used for emergency services.

_____

_____

_____ 12. Vehicle mounted generators can be powered by PTOs.

_____

_____

_____ 13. The most commonly used extrication tools are electrically powered.

_____

_____

Multiple Choice

C. Write the letter of the best answer on the blank before each statement.

____ 1. What NFPA standard contains the requirements for pumper design?
   a. 1751
   b. 1801
   c. 1901
   d. 2001

____ 2. According to the NFPA standard, the minimum pump capacity for pumpers is ____.
   a. 500 gpm *(2 000 L/min)*
   b. 750 gpm *(3 000 L/min)*
   c. 1,000 gpm *(4 000 L/min)*
   d. 1,250 gpm *(5 000 L/min)*

____ 3. What is usually the maximum pumping capacity of municipal fire department pumpers?
   a. 1,000 gpm *(4 000 L/min)*
   b. 1,500 gpm *(6 000 L/min)*
   c. 2,000 gpm *(8 000 L/min)*
   d. 2,500 gpm *(10 000 L/min)*

____ 4. Which of the following statements about pumpers is true?
   a. The NFPA standard requires fire department pumpers to carry portable water tanks.
   b. The equipment on any particular pumper should not exceed the NFPA standard's specifications.
   c. All equipment carried on fire department pumpers should be associated directly with fire stream/water supply functions.
   d. Rescue pumpers typically have more storage space than standard pumpers.

____ 5. Most industrial foam pumpers are primarily intended to ____.
   a. Flow foam solution on Class B fires
   b. Flow plain water on Class A fires
   c. Flow foam solution on Class A fires
   d. Flow plain water on Class B fires

____ 6. Which of the following is the NFPA Standard for Mobile Foam Apparatus?
   a. 12B
   b. 303
   c. 11C
   d. F10

____ 7. Which of the following statements about foam pumpers is true?
   a. Some foam pumpers have foam proportioning systems capable of exceeding the pumping capacity of the apparatus by itself.
   b. Foam pumpers must be mounted on custom built chassis.
   c. Municipal department foam pumpers equipped with Class B foam systems typically can handle a major flammable liquid fire or spill.
   d. Class A foam systems on municipal apparatus must be the low-energy type.

____ 8. The most common size foam concentrate tanks for municipal fire apparatus range from ____.

a. 50 to 200 gallons *(200 L to 800 L)*

b. 20 to 100 gallons *(80 L to 400 L)*

c. 25 to 75 gallons *(100 L to 300 L)*

d. 100 to 150 gallons *(400 L to 600 L)*

____ 9. Which of the following statements about minipumpers is true?

a. Most are mounted on a three-quarter ton chassis.

b. They are too small to set up a master stream.

c. They are routinely equipped with pumps as large as 1,000 gpm *(4 000 L/min)*.

d. Some carry advanced rescue and extrication equipment.

____ 10. For what type of fire would a midipumper be best suited?

a. House fire

b. Large Class B fire

c. Dumpster fire

d. Large wildland fire

____ 11. Why should a mobile water supply apparatus be similar in size to other tenders (tankers) in the area?

a. Oncoming apparatus can pass more safely.

b. Drivers can readily trade apparatus.

c. Water shuttles will flow more easily.

d. Financial resources will be allotted equally.

____ 12. According to the NFPA standard, what minimum number of gallons *(liters)* must an apparatus carry to be considered a mobile water supply apparatus?

a. 1,500 gallons *(6 000 L)*

b. 1, 000 gallons *(3 785 L)*

c. 750 gallons *(3 000 L)*

d. 500 gallons *(2 000 L)*

____ 13. Generally, what is the maximum tank capacity for vehicles with a single rear axle?

a. 1,500 gallons *(6 000 L)*

b. 1, 000 gallons *(3 785 L)*

c. 750 gallons *(3 000 L)*

d. 500 gallons *(2 000 L)*

____ 14. What is the largest tank size on most wildland fire apparatus?

a. 250 gallons *(1 000 L)*

b. 500 gallons *(2 000 L)*

c. 750 gallons *(3 000 L)*

d. 1,000 gallons *(4 000 L)*

____ 15. Which of the following statements is true?

a. Class A foam agents are ineffective in fighting wildland fires.

b. NFPA 1500 prohibits using high-energy foam systems on wildland fire apparatus.

c. A wildland fire apparatus may be top heavy if it has baffled tanks.

d. Booster tanks for wildland fire apparatus may exceed 1,000 gallons *(4 000 L)*.

____ 16. Requirements for ARFF apparatus are contained in NFPA ____.

a. 414
b. 1144
c. 1901
d. 441

____ 17. Which of the following is ***not*** an NFPA classification for ARFF apparatus?

a. Major fire fighting vehicles
b. Rapid intervention vehicles
c. Combined agent vehicles
d. Containment vehicles

____ 18. Airports outside the United States follow requirements in ____.

a. 14 CFR Part 139
b. UNAV Chapter 12
c. (FAA) AC 150/5220-10
d. ICAO Annex 14

____ 19. The two fire fighting operations for which fire boats are best suited are ____.

a. Pumping through large master stream devices and generating foam
b. Shuttling Class D extinguishing agents and generating foam
c. Pumping through large master stream devices and relaying water to land-based apparatus
d. Generating foam and relaying water to land-based apparatus

____ 20. Which of the following features distinguishes a quint from a quad?

a. Fire pump
b. Aerial device
c. Fire hose
d. Ground ladders

____ 21. To be considered a true quint, an aerial apparatus must have a pump with a capacity of at least ____.

a. 250 gpm *(1 000 L/min)*
b. 500 gpm *(2 000 L/min)*
c. 750 gpm *(3 000 L/min)*
d. 1,000 gpm *(4 000 L/min)*

____ 22. Generally rescue apparatus equipped with a fire pump carry ____ of water or less.

a. 250 gallons *(1 000 L)*
b. 500 gallons *(2 000 L)*
c. 750 gallons *(3 000 L)*
d. 1000 gallons *(4 000 L)*

____ 23. Inverters on pumpers are generally capable of providing approximately ____ of electric power.

a. 1,200 watts *(1.2 kW)*
b. 1,300 watts *(1.3 kW)*
c. 1,400 watts *(1.4 kW)*
d. 1,500 watts *(1.5 kW)*

____ 24. The largest portable generators can provide a maximum of ____ of electric power.

a. 2,500 watts *(2.5 kW)*
b. 5,000 watts *(5 kW)*
c. 7,500 watts *(7.5 kW)*
d. 10,000 watts *(10 kW)*

____ 25. Generators on rescue vehicles commonly have maximum capacities of up to _____.

a. 35,000 watts *(35 kW)*
b. 40,000 watts *(40 kW)*
c. 45,000 watts *(45 kW)*
d. 50,000 watts *(50 kW)*

____ 26. Which of the following statements about portable lighting equipment is true?

a. They must be supplied by a cord from the power plant.
b. They generally range from 250 to 500 watts.
c. They usually have handles and large bases.
d. They must be held by hand.

____ 27. Which of the following statements is true?

a. Fixed lighting is usually used to spotlight areas that portable lights cannot reach.
b. Banks of fixed lights generally have a capacity of 1,000 to 2,000 watts per light.
c. The most common electrical power distribution cable is a 12-gauge, 3-wire type.
d. Emergency scene electrical power distribution should ***not*** use junction boxes.

28. Which of the following is ***not*** one of the four basic types of powered hydraulic tools used by the rescue service?

a. Extension ram
b. Jack hammer
c. Shears
d. Spreader

____ 29. Most pumps are ***not*** capable of supplying full power to a hydraulic tool when the hose length between the pump and the tool exceeds ____.

a. 100 feet *(30 m)*
b. 75 feet *(23 m)*
c. 50 feet *(15 m)*
d. 30 feet *(9 m)*

____ 30. According to the NIIMS Incident Command System, what is the minimum water tank capacity in gallons *(liters)* for a Type 5 pumper?

a. 70 *(280)*
b. 50 *(200)*
c. 750 *(3 000)*
d. 200 *(800)*

____ 31. According to the NIIMS Incident Command System, what is the minimum pump capacity in GPM *(L/min)* for a Type 1 water tender?

a. 300 *(1 200)*
b. 5,000 *(20 000)*
c. 200 *(800)*
d. 1,000 *(4 000)*

Identify

D. Identify the following abbreviations associated with types of fire apparatus equipped with a fire pump. Write the correct interpretation before each.

_______________ 1. GVW

_______________ 2. PTO

_______________ 3. ARFF

_______________ 4. CFR

_______________ 5. ICAO

_______________ 6. FAA

_______________ 7. RIV

_______________ 8. ICS

_______________ 9. NWCG

Chapter 3 Introduction to Apparatus Inspection and Maintenance

Matching

A. Match to their definitions terms associated with apparatus inspection and maintenance. Write the appropriate letters on the blanks.

____ 1. Keeping an apparatus in a state of usefulness or readiness

____ 2. Restoring or replacing that which has become inoperable

____ 3. Turns on lights at specified intervals during startup to prevent putting too much demand on the electrical system

____ 4. Shuts down less important equipment to prevent putting too much demand on the electrical system

____ 5. Distance the clutch pedal must be pushed before the throw-out bearing actually contacts the clutch release fingers

a. Free play
b. Maintenance
c. Power takeoff
d. Repair
e. Load monitor
f. Load sequencer

True/False

B. Write *True* or *False* before each of the following statements. Correct those statements that are false.

________ 1. Regularly performing routine maintenance checks can prevent most apparatus or equipment failures.

____

____

________ 2. Older apparatus require gentler cleaning procedures than do newer apparatus.

____

____

_______ 3. Dirt can cause deterioration of engine compartment coverings.

____

____

_______ 4. The driver/operator should not allow the records of previous inspections to influence the current inspection.

____

____

_______ 5. The driver/operator should check each lug nut's tightness by hand.

____

____

_______ 6. Acceptable tire tread depth varies according to state or provincial inspection requirements.

____

____

_______ 7. Cracked headlight lenses need not be replaced until the bulb burns out.

____

____

_______ 8. A second firefighter in the cab should test the operation of audible warning devices while the first firefighter checks the headlights.

____

____

_______ 9. Pump intake valves at the front of the vehicle should be fully open when not in use.

____

____

_______ 10. Equipment in compartments should not be inspected at the time of the vehicle inspection.

____

____

_________ 11. NFPA 1901 requires that a new apparatus must be brought to a complete stop from a speed of 20 mph *(32 km/h)* in a distance not to exceed 35 feet *(10.7 m)*.

____

____

_________ 12. Apparatus with air brakes are to be equipped with an air pressure protection valve that prevents the air horns from operating when air reservoir pressure drops below 80 psi *(552 kPa)*.

____

____

_________ 13. If the radiator coolant level is low, the driver/operator should add water to the system.

____

____

_________ 14. Most modern vehicle batteries do not require internal inspection.

____

____

_________ 15. Driver/operators should perform posttrip inspections after operating at a fire for a long period or after an unusually long road trip.

____

____

Multiple Choice

C. Write the letter of the best answer on the blank before each statement.

____ 1. The driver/operator should be able to perform ____.

a. Basic maintenance
b. Detailed maintenance
c. Minor repairs
d. Major repairs

____ 2. Which of the following is ***not*** subject to SOPs regarding apparatus maintenance?

a. Who performs maintenance functions
b. When maintenance functions are to be performed
c. How the maintenance process is documented
d. How much is budgeted for maintenance tools, supplies, and work hours

____ 3. Typically, how often do career fire departments require driver/operators to perform apparatus inspections and maintenance checks?
   a. At the beginning of each day
   b. At the beginning of each tour of duty
   c. At the beginning of each work week
   d. At the beginning of each month

____ 4. Which of the following is least likely to be a function of maintenance and inspection records?
   a. Warranty claims
   b. Accident investigations
   c. Driver/operator evaluations
   d. Apparatus purchases

____ 5. Which of the following is ***not*** a benefit of vehicle and equipment cleanliness?
   a. Removal of lubricants
   b. Public relations
   c. Easier inspection
   d. Longer vehicle life

____ 6. Which of the following guidelines for washing the exterior of a new apparatus is correct?
   a. Use a garden hose with a nozzle or a pressure washer.
   b. Use hot, sudsy water.
   c. Rinse the vehicle before shampooing.
   d. Remove grease or tar with gasoline.

____ 7. Which of the following is a valid guideline for cleaning automotive glass?
   a. Use cool, clear water
   b. Remove deposits with a putty knife
   c. Polish glass with a dry cloth
   d. Apply cleaner with damp paper towels or rags

____ 8. Which of the following is a valid guideline for waxing apparatus?
   a. Newer apparatus require more frequent waxing than do older apparatus.
   b. Wax can damage clear-coat finishes.
   c. Waxes should not be applied until the apparatus is at least 12 months old.
   d. Mechanical buffers should not be used on older vehicles.

____ 9. What is the final step in the circle or walk-around inspection method?
   a. Inspect driver's door on the cab.
   b. Perform functional check of apparatus systems.
   c. Check engine compartment.
   d. Inspect underneath apparatus.

____ 10. Which of the following is a true statement about approaching the vehicle during the inspection process?
   a. Apparatus should be expected to lean to one side.
   b. Spots beneath the vehicle are normal.
   c. Chocking the wheels is not necessary when the apparatus is parked on a level surface.
   d. The inspection process should begin as the driver/operator approaches the vehicle.

____ 11. The proper order for left- and right-front side inspection is ____.
   a. Doors, equipment in rear of cab, fuel tank, wheel and tire
   b. Wheel and tire, fuel tank, doors, equipment in rear of cab
   c. Doors, fuel tank, wheel and tire, equipment in rear of cab
   d. Fuel tank, wheel and tire, doors, equipment in rear of cab

____ 12. What should the driver/operator note first when inspecting the front of the vehicle?
   a. Significant body damage
   b. Condition of windshield
   c. Condition of lights
   d. Visible damage to audible warning devices

____ 13. The correct order for inspecting the front of the apparatus is ____.
   a. Front end, right-front side, left-front side
   b. Right-front side, left-front side, front end
   c. Left-front side, front end, right-front side
   d. Left-front side, right-front side, front end

____ 14. What part of the apparatus should rear-side inspections cover?
   a. Front of cab to tailboard
   b. Front of truck to tailboard
   c. Rear of cab to tailboard
   d. Rear of cab to rear axle

____ 15. The process of automatically shutting down less important electrical equipment to prevent an overload is called ____.
   a. Demand dropping
   b. Load shedding
   c. Work reduction
   d. Watt watching

____ 16. In general, steering wheel play should be no more than about ____ degrees in either direction.
   a. 7
   b. 10
   c. 12
   d. 15

____ 17. NFPA 1901 requires the parking brake to hold the apparatus in place on grades of up to ____ percent.
   a. 7
   b. 15
   c. 25
   d. 20

____ 18. How soon after starting the apparatus should air pressure be sufficient to allow vehicle operation?

a. 60 seconds
b. 90 seconds
c. 120 seconds
d. 150 seconds

____ 19. Properly functioning brakes should stop an emergency vehicle traveling at 5 mph *(10 km/h)* within ____ of the point where the brake pedal was pushed.

a. 20 feet *(6 m)*
b. 25 feet *(8 m)*
c. 30 feet *(9 m)*
d. 50 feet *(15 m)*

____ 20. With the exception of the ____, fluid level checks should be done while the engine is turned off.

a. Engine oil
b. Automatic transmission fluid
c. Radiator coolant
d. Brake fluid

____ 21. A lubricant's SAE number indicates its ____.

a. Corrosion protection
b. Tendency to foam
c. Sludge factor
d. Viscosity

____ 22. Batteries produce ____ when being charged.

a. Static electricity
b. Hydrogen gas
c. Water vapors
d. Hydrochloric acid

Identify

D. Identify the following abbreviations associated with apparatus inspection and maintenance. Write the correct interpretation before each.

____ 1. CDL

____ 2. ABS

____ 3. SAE

E. Identify daily and weekly pump maintenance functions. Write a *D* before those functions that should be performed daily. Write a *W* before those functions that should be performed weekly.

____ 1. Flush pump with clear water.
____ 2. Operate pump drive control.
____ 3. Make sure auxiliary fuel tank is full.
____ 4. Check and clean intake strainers.
____ 5. Make sure pump and booster lines are completely drained of water.

____ 6. Operate controls to check or inspect fire pump.
____ 7. Check pump gear box oil level.
____ 8. Check underside of apparatus and inside of compartments for water or foam leaks.
____ 9. Recalibrate flow meter.
____ 10. Test roof and bumper turrets.

List

F. List five items that should be checked during left- and right-front side inspections, if present.

1. ______________________________
2. ______________________________
3. ______________________________
4. ______________________________
5. ______________________________

G. List ten items that should be checked during front end inspections, if present.

1. ______________ 6. ______________
2. ______________ 7. ______________
3. ______________ 8. ______________
4. ______________ 9. ______________
5. ______________ 10. ______________

H. List five items that should be checked during left- and right-rear side inspections, if present.

1. ______________________________
2. ______________________________
3. ______________________________
4. ______________________________
5. ______________________________

I. **List five items that should be checked during rear inspections, if present.**

1. ______
2. ______
3. ______
4. ______
5. ______

J. **List five types of gauges that should be checked during inspections, if present.**

1. ______
2. ______
3. ______
4. ______
5. ______

K. **List five controls located in the cab that should be checked during inspections.**

1. ______
2. ______
3. ______
4. ______
5. ______

L. **List ten items that should be checked during engine compartment inspections.**

1. ______
2. ______
3. ______
4. ______
5. ______
6. ______
7. ______
8. ______
9. ______
10. ______

Chapter 4 Operating Emergency Vehicles

Matching

A. Match to their definitions terms associated with operating emergency vehicles. Write the appropriate letters on the blanks.

____ 1. Scanning far enough ahead to ensure the driver/operator will have enough time to take any appropriate action

____ 2. Sum of driver/operator reaction distance and vehicle braking distance

____ 3. Distance a vehicle travels while a driver is transferring the foot from the accelerator to the brake pedal after seeing the need to stop

____ 4. Distance the vehicle travels from the time the brakes are applied until the apparatus comes to a complete stop

____ 5. Reduces the air pressure on the front steering axle by 50 percent when in the "slippery road" position

____ 6. Allows for a difference in speed between tandem rear axles while providing pulling power from each axle

____ 7. Improves traction on slippery roads by reducing drive wheel overspin

____ 8. Increases traction on extra soft surfaces

____ 9. System in which strobe lights on the apparatus signal sensors on traffic lights

____ 10. System in which traffic signals respond to emergency vehicle sirens

____ 11. Tests the driver/operator's ability to move the vehicle backward within a restricted area

____ 12. Simulates maneuvering around parked and stopped vehicles and tight corners

____ 13. Measures a driver/operator's ability to steer in a straight line, judge distances from wheel to object, and stop at a finish line

a. Snow and mud switch
b. Braking distance
c. Diminishing-clearance exercise
d. Front-brake limiting valve
e. Confined space turnaround
f. Visual lead time
g. Alley dock exercise
h. Total stopping distance
i. Serpentine course
j. Opticom
k. Reaction distance
l. Automatic traction control
m. SONEM 2000
n. Interaxle differential lock

True/False

B. Write *True* or *False* before each of the following statements. Correct those statements that are false.

_______ 1. Studies in Pennsylvania and New York indicate that most collisions involving EMS vehicles occur in broad daylight on dry roads.

____

____

_______ 2. Inattentiveness and panic are common causes of reckless driving by the public.

____

____

_______ 3. Homebuilt vehicles are less likely to have design problems than those built by a fire apparatus manufacturer.

____

____

_______ 4. The driver/operator should wait until all occupants are within the cab, seated, and wearing seat belts before starting the apparatus.

____

____

_______ 5. The driver/operator should wait until all occupants are within the cab, seated, and wearing seat belts before putting the apparatus in motion.

____

____

_______ 6. Most departments have standard operating procedures for times when the apparatus may be forced to idle for an extended time.

____

____

False 7. If they are equipped with safety bars or gates and seat belts, jump seats are acceptable substitutes for fully enclosed seats.

____

____

_____ 8. Fire apparatus on an emergency response must come to a complete stop before proceeding through a red light or stop sign.

_____

_____

False 9. The extra weight carried on most fire apparatus helps to prevent skids and rollover.

_____

_____

_____ 10. The most common causes of skids involve brake failures.

_____

_____

_____ 11. In a vehicle equipped with an ABS, the driver/operator should maintain a steady pressure on the brake pedal rather than pumping it if the vehicle goes into a skid.

_____

_____

True IFSTA recommends placing a front-brake limiting valve in the dry-road position and disconnecting it.

_____

_____

_____ 13. A siren operating on an emergency vehicle moving at 60 mph *(97 km/h)* is audible 300 feet *(90 m)* in front of the vehicle.

_____

_____

TRUE 14. When firefighters must be in the street, it is desirable to turn off some of the warning lights on the apparatus once it is parked.

_____

_____

_____ 15. NFPA 1002 provides specific directions on testing driver/operator candidates.

_____

_____

4

Multiple Choice

C. Write the letter of the best answer on the blank before each statement.

____ 1. What percentage of firefighter injuries and deaths are caused by vehicle collisions while responding to or returning from emergency calls?

a. 10–15
b. 15–20
c. 20–25
d. 25–30

____ 2. Approximately how many civilians are killed each year in emergency-apparatus related collisions?

a. 10
b. 25
c. 50
d. 100

D 3. Which of the following is ***not*** one of the five basic causes of fire apparatus collisions?

a. Improper apparatus backing
b. Excessive apparatus speed
c. Poor apparatus design or maintenance
d. Adverse road conditions

A 4. Which of the following statements about air brake systems is true?

a. They activate slower than hydraulic brakes but stop the vehicle faster.
b. They activate faster than hydraulic brakes and stop the vehicle faster.
c. They activate slower than hydraulic brakes and stop the vehicle slower.
d. They activate faster than hydraulic brakes but stop the vehicle slower.

____ 5. Which of the following statements is true?

a. Driver/operator candidates should drive apparatus under emergency conditions during their training.
b. Driver/operators with a good driving record can be confident that they will not have an accident.
c. An apparatus with a full water tank is more likely to skid than one with an empty water tank.
d. In about 40 percent of all collisions, the driver/operator is not aware of a problem until it is too late to correct it.

____ 6. Which of the following vehicle types has had a particularly high incidence of serious collisions?

a. Factory built ladder truck
b. Factory built tanker
c. Homebuilt aerial vehicle
d. Homebuilt tanker

____ 7. Which of the following vehicle systems is particularly likely to lead to a collision if it fails?

a. Braking
b. Electrical
c. Power train
d. Suspension

C ___ 8. Which of the following statements about driving regulations is true?

a. Unless otherwise specified, emergency vehicle driver/operators are exempt from traffic laws and regulations.
b. Most driving regulations pertain to dry, clear roads during daylight conditions.
c. Emergency vehicles are generally exempt from laws that require vehicles to stop for school buses with flashing signal lights.
d. If a driver/operator's negligence contributes to a collision, the fire department cannot be held responsible.

C 9. What should a driver/operator do first when starting an apparatus?

a. Sound an audible warning device.
b. Disconnect ground shore lines.
c. Know where the apparatus is going.
d. Turn on vehicle batteries.

___ 10. Which of the following statements about driving a manual transmission apparatus is true?

a. The driver/operator should start the apparatus in high gear when it is going downhill.
b. The driver/operator should select a lower gear when driving downhill.
c. When turning corners, the driver operator should shift into a lower gear after entering the intersection
d. The faster an engine goes, the more power it develops.

D 11. Which of the following statements about driving an automatic transmission apparatus is true?

a. If the accelerator is only partially depressed, the transmission automatically shifts near the governed speed of the engine.
b. Automatic transmissions override the driver/operator's gear selections.
c. Shifting at high engine speeds increases power and energy efficiency.
d. Some shifters have a normal driving (D) designation similar to that found on smaller passenger vehicles.

C 12. Which of the following gear selections would be most appropriate for operating a manual transmission apparatus in heavy traffic?

a. 5
b. 4
c. 3
d. 2

D 13. Which of the following gear selections would be most appropriate for driving an automatic transmission apparatus on moderate grades and for over-the-road operation at moderate speeds?

a. Overdrive
b. 2
c. D
d. 1–4

14. Which of the following statements about cruising is true?
    a. Driver/operators should stay in a lower gear as long as possible before shifting to a higher gear.
    b. When ascending a steep grade, momentary lugging is unavoidable.
    c. Overthrottling occurs in diesel engines when less fuel is being injected than can be burned.
    d. The driver/operator should choose a gear that cruises the engine slightly above the manufacturer's recommended rpm.

____ 15. Some apparatus employ ____ that assist in braking.
    a. Engine brakes
    b. Governors
    c. Tachometers
    d. Torque converters

16. When should the clutch be disengaged during braking?
    a. Shortly before applying the brakes
    b. Simultaneously to applying the brakes
    c. Shortly after applying the brakes
    d. Shortly before the apparatus comes to a full stop

____ 17. Which of the following is ***not*** a result of long idling periods?
    a. Reduced engine wear
    b. Carbon buildup
    c. Misfiring
    d. Damage to turbocharger shaft seals

18. When the engine must be left to idle for an extended time, it should be set at ____ rpm.
    a. 300–500
    b. 500–700
    c. 700–900
    d. 900–1,100

19. Generally, how long should a hot engine idle before shutting down?
    a. Less than 3 minutes
    b. 3 to 5 minutes
    c. 5 to 10 minutes
    d. 10 minutes or more

____ 20. Which of the following is the best way of looking at emergency driving?
    a. The driver/operator has to drive more aggressively than the public at large.
    b. Driver/operators have the legal right of way and occasionally must demand it.
    c. Driver/operators should drive as they ordinarily would and take advantage of the room that clears for them on the road.
    d. The public relations of arriving at an emergency outweigh the public relations of driving cautiously in traffic.

21. Which of the following does the NFPA ***not*** recognize as an exception to its requirement that all riders on an apparatus should be seated within the cab or body and wearing seat belts?
    a. Providing patient care in the back of an ambulance
    b. Loading hose back onto an apparatus
    c. When doing a pump and roll attack on a wildland fire
    d. Training personnel to drive the tiller or a tractor-drawn aerial apparatus

22. Which of the following statements about hose loading procedures is ***not*** true?
    a. At least one safety observer must have complete visual contact with the hose-loading operation and visual and voice communications with the driver/operator.
    b. The area must be closed to other vehicular traffic.
    c. The apparatus must be driven at no more than 5 mph *(8 km/h)*.
    d. Any member standing on the tailboard must use a safety strap.

23. Which of the following statements about backing an apparatus is true?
    a. It is normally safer and sometimes quicker to drive around the block and start again.
    b. All fire stations should be designed with extra-wide apparatus bays to make backing easier.
    c. When two spotters help with backing, both should communicate directly with the driver/operator.
    d. If spotters are not available or cannot be seen, the driver/operator should back the apparatus with extreme caution.

24. Where is the most likely place for a collision involving an emergency vehicle?
    a. Access to street from fire department driveway
    b. Intersection
    c. Emergency scene
    d. Highway entrance ramp

25. Which of the following is ***not*** a defensive driving technique?
    a. Aim high in steering.
    b. Get the big picture.
    c. Keep your eyes stationary.
    d. Leave yourself an "out."

____ 26. Which of the following does ***not*** directly affect braking distance?
    a. Age of vehicle
    b. Condition of the driving surface
    c. Speed being traveled
    d. Weight of the vehicle

____ 27. What is the approximate braking distance (feet) of a light two-axle truck going 35 mph on dry, level pavement?
a. 39
b. 92
c. 58
d. 45

____ 28. What is the average driver reaction distance (feet) at 55 mph?
a. 165
b. 66
c. 61
d. 226

____ 29. What is the average total stopping distance (meters) for a three axle truck and trailer going 64 km/h on dry, level pavement?
a. 60.7
b. 63.7
c. 62.5
d. 75.9

____ 30. What should a driver/operator do if a vehicle that is ***not*** equipped with an ABS goes into a skid?
a. Gradually apply power to the wheels once the skid is controllable
b. Maintain steady pressure on the brake pedal
c. Turn the apparatus steering wheel so that the front wheels face away from the direction of the skid
d. Release the clutch

____ 31. What should a driver/operator do when behind a slower moving vehicle?
a. Pass civilian vehicles on the right.
b. Pass other emergency apparatus on the right.
c. Travel on the outermost lane on multilane roads.
d. Avoid passing vehicles that do not pull over to yield the right-of-way.

____ 32. Which of the following is ***not*** usually one of the first areas to become slippery?
a. Bridge surfaces
b. Southern slopes of hills
c. Shaded spots
d. Areas where snow is blowing across the roadway

____ 33. How much more stopping distance does an apparatus need on snow and ice than on dry concrete?
a. 2–5 times
b. 2–10 times
c. 3–10 times
d. 3–15 times

____ 34. Which of the following is the surest way to capture a driver's attention?
a. Bells
b. Long blasts with the air-horns
c. Siren that maintains a steady frequency
d. Short bursts with the air-horns

____ 35. Which of the following statements is true?

a. Most jurisdictions consider water shutoffs to be emergency calls.
b. Warning devices are usually unnecessary on limited-access highways and turnpikes.
c. Units responding along the same route should travel a maximum of 300 feet *(90 m)* apart.
d. White lights are ineffective during daylight hours.

____ 36. All fire apparatus training and testing should follow the requirements contained in ____, *Standard for a Fire Service Vehicle Operations Training Program.*

a. NFPA 1451
b. USDOT 2.009
c. OSHA (EMS) 55
d. NFPA 1002

D 37. The ____ is an acceptable alternative to the alley dock exercise.

a. Confined space turnaround
b. Written test
c. Parallel parking test
d. Apparatus station parking maneuver

B 38. How far apart should markers be for the serpentine exercise?

a. 30 feet to 50 feet *(9 m to 16 m)*
a. 30 feet to 38 feet *(9 m to 12 m)*
a. 38 feet to 50 feet *(12 m to 16 m)*
a. 50 feet to 58 feet *(16 m to 19 m)*

D 39. How many direction changes does the confined space turnaround exercise allow?

a. 2
b. 6
c. 10
d. Unlimited

C 40. How close to the finish line must the driver/operator stop the front bumper in the diminishing-clearance test?

a. 12 inches *(300 mm)*
b. 9 inches *(225 mm)*
c. 6 inches *(150 mm)*
d. 3 inches *(75 mm)*

Identify

D. Identify the following abbreviations associated with operating emergency vehicles. Write the correct interpretation before each.

Privately Owned Vehicle 1. POV

Emergency Medical System 2. EMS

_______________ 3. ABS

automatic traction control 4. ATC

E. Identify NFPA 1002 requirements for road tests. Write an *X* before each correct requirement below.

___ 1. Five left and five right turns

X 2. A straight urban business street or two-lane rural road at least one mile long

X 3. A railroad crossing

___ 4. Two curves; one left and one right

___ 5. A section of limited-access highway that includes a conventional on-ramp and off-ramp and is long enough to allow at least four lane changes

X 6. An upgrade that is steep enough and long enough to require gear changing to maintain speed

X 7. One underpass, low-clearance bridge

Chapter 5

76

A

1 H
2 ~~G~~ B (88)
3 ~~L~~ G (89)
4 D ~~[illegible] (89)~~
5 A
6 F
7 J
8 E
9 I

B

1 T
2 T
3 F upwind
4 T
~~5 T~~ F
6 F
~~7 T~~ F
8 T
9 F
10 T
11 T
12 T
13 F
14 T
15 F
16 T
~~17 T~~ F
18 F

C

1 A
~~2 D~~ A 77
~~3 C~~ D 77,8
4 C
5 B
~~6 C~~ B 80
7 A
~~8 B~~ C 81
9 C
~~10 A~~ B Not needed
11 metric
12 D
13 A
~~14 A~~ ?C 86
15 d
16 B
17 B
18 C
~~19 C~~ A 89
20 B
21 C
22 D
~~23 A~~ D
24 C
25 A
26 C
27 B
28 D
~~29 100S C~~ B 96
30 B
31 A
32 D
33 A
34 C
35 C
~~36 A~~ B10

Chapter 5 Positioning Apparatus

Matching

A. Match to their definitions terms associated with positioning apparatus. Write the appropriate letters on the blanks.

____	1. Length of strainer-equipped piping extending from a static water supply source to a hose connection on the shore	a. Cascade system
B	2. Two pumpers connected intake-to-intake on one hydrant	b. Dual pumping
G	3. Two pumpers connected in series to one hydrant	c. Interlock
____	4. Boundary between wildland and structural development	d. Wildland/urban interface
____	5. Large breathing-air cylinders connected in banks	e. Level I staging
____	6. Engine-driven appliance that purifies and compresses air	f. Breathing air compressor
____	7. Non-transport EMS vehicle	g. Tandem pumping
____	8. Used whenever two companies performing like functions are dispatched to an emergency scene	h. Dry hydrant
____	9. Used when incidents require mutual aid or result in multiple alarms	i. Level II staging
		j. Paramedic/quick responder unit

True/False

B. Write *True* or *False* before each of the following statements. Correct those statements that are false.

________ 1. There is no one set rule for positioning pumpers supplying attack lines on the fireground.

____

____

_______ 2. Standard operating procedures and the judgment of the responsible officer or driver/operator should be the deciding factors when committing/positioning the apparatus.

____

____

_______ 3. Whenever possible the driver/operator should position the apparatus downwind of the incident.

____

____

_______ 4. The driver/operator should always position the apparatus where it can make an easy retreat.

____

____

F 5. Fire hydrants are not usually located very close to the fire department connection.

____

____

_______ 6. Pumpers supporting a fire department connection always take positioning priority over all other apparatus.

____

____

F 7. Drafting operations are common in rural areas but are much more difficult to perform at urban incidents.

____

____

_______ 8. Suction hose may be laid on top of a ground ladder to keep the strainer off the bottom of the water source.

____

____

_______ 9. IFSTA recommends using hard suction hose to connect a pumper to a fire hydrant.

____

____

_________ 10. If the front wheels of an apparatus are turned to a 45-degree angle, the driver/operator can easily adjust the distance to or from the hydrant by moving the unit forward or backward.

____

____

_________ 11. A bell reducer or suction siamese fitting is a better arrangement than separate connections to two 2½-inch *(65-mm)* outlets.

____

____

_________ 12. A pumper may use both a large diameter intake and smaller hoselines from one exceptionally strong hydrant.

____

____

_________ 13. Most wildland fire fighting SOPs require the driver/operator not to drive into areas where smoke obscures visibility.

____

____

_________ 14. In general, the positioning of rescue apparatus is not as critical as that of pumping and aerial apparatus.

____

____

_________ 15. In level II staging, the incident commander gives orders directly to the staged units.

____

____

_________ 16. First-arriving apparatus should never drive directly into a hazardous materials scene without first attempting to identify the material that is involved.

____

____

F 17. Workers are required to wear protective clothing in the warm zone at a hazardous material incident.

____

____

____ 18. The command post and staging area are located within the warm zone at a hazardous material incident.

____

____

Multiple Choice

C. Write the letter of the best answer on the blank before each statement.

____ 1. Generally the best place to park an apparatus at an incident where no fire conditions are evident is ____.

a. Near the main entrance
b. Near the standpipe connection
c. Near the closest hydrant
d. Near a secondary entrance

A 2. Which of the following is the first priority in determining apparatus placement?

a. Rescue
b. Water supply
c. Exposures
d. Method of attack

D 3. Which of the following statements is true?

a. Apparatus should always be positioned to best protect the original fire building.
b. Placement of the initial apparatus is less important than placement of later arriving apparatus.
c. Water supply is always a major consideration in positioning the apparatus.
d. Apparatus positioned in long narrow driveways should lay their own supply line.

____ 4. When is it best to position the apparatus downhill from the incident?

a. Incident that requires hoselines
b. Hazardous materials incident
c. Wildland fire
d. Vehicle fire

____ 5. The best position for the supply hose is ____.

a. The middle of the street
b. The side of the street
c. The middle of the incoming traffic lane
d. The middle of the outgoing traffic lane

6. The collapse zone should be at least ____ time(s) the height of the building.
   a. One-half
   b. One
   c. One and one-half
   d. Two

7. Ornamental stars usually indicate ____.
   a. Reinforcement ties to hold unstable walls
   b. Well maintained structures
   c. Older construction styles
   d. Reliable construction methods

8. If falling debris becomes a hazard and the driver/operator cannot reposition the apparatus, what should be done?
   a. Retreat with the apparatus and "write off" the structure.
   b. Abandon the apparatus.
   c. Cover the apparatus with tarps or salvage covers.
   d. Erect approved debris barriers around the bottom of the apparatus.

9. Which of the following statements is true?
   a. In most cases it is best to give the pumper the optimum operating position.
   b. Most pumpers have fewer than 1,000 feet *(300 m)* of hose.
   c. If a building is higher than five stories, the aerial device should be positioned on the inside (closest to the building) and the pumper on the outside.
   d. Pumpers providing water for elevated stream operations should be positioned as close to the hydrant as possible.

10. What is the maximum discharge (U.S.) of a 1,000 gpm *(4 000 L/min)* rated pump from two sections of 5-inch *(125 mm)* intake hose with a lift of 12 feet *(3.6 m)*?
    a. 770 gpm *(2 195 L/min)*
    b. 935 gpm *(3 539 L/min)*
    c. 1,000 gpm *(3 785 L/min)*
    d. 1,020 gpm *(3 861 L/min)*

11. What is the minimum pump capacity (metric) needed to discharge 1 287 L/min from three sections of 115 mm intake hose with a lift of 6.7 m?
    a. 1 000 L/min
    b. 2 000 L/min
    c. 3 000 L/min
    d. 4 000 L/min

12. Why do jurisdictions require the driver/operator to place gated valves on a hydrant's small-diameter discharges when making a connection to the large-diameter discharge?
    a. For pressure relief
    b. For back flushing hoselines
    c. For vacuum release
    d. For additional hose connections

____ 13. The preferred type of hose for connection to a fire hydrant is ____.

a. Large-diameter soft intake hose
b. Large-diameter hard suction hose
c. Small-diameter soft intake hose
d. Small-diameter hard suction hose

____ 14. Intake hose sections are commonly ____ long.

a. 10–20 feet *(3 m to 6 m)*
b. 20–25 feet *(6 m to 7.5 m)*
c. 10–50 feet *(3 m to 15 m)*
d. 25–50 feet *(7.5 m to 15 m)*

____ 15. To avoid kinks in soft-sleeve hose, the driver/operator should ____.

a. Stop the pumper with the side intake in line with the hydrant.
b. Put two half-twists in the hose.
c. Put two full twists in hose if both ends are equipped with Storz couplings.
d. Stop the pumper with the side intake a few feet *(meters)* short of being in line with the hydrant.

____ 16. When using front or rear intake connections, the vehicle should be aimed toward the hydrant at an angle of ____.

a. 60 degrees or less
b. 45 degrees or less
c. 45 degrees or more
d. 60 degrees or more

____ 17. What is the main disadvantage of connecting to a hydrant with one or two 2½-inch *(65-mm)* outlets?

a. Shorter hose lengths
b. Limited water supply
c. Weak construction
d. Increased maneuvering time

____ 18. One result of dual pumping is ____.

a. Less efficient use of available water
b. Longer hose lays
c. Closer apparatus groupings
d. Longer times for placing additional hoselines

____ 19. Which of the following is ***not*** a common use for tandem pumping operations?

a. Wildland fire
b. Support supply line when attack pumper is close to hydrant
c. High-rise sprinkler
d. Standpipe

____ 20. The highest priority for most wildland fire fighting operations is ____.

a. Wildlife preservation
b. Structural preservation
c. Air-quality preservation
d. Vegetation preservation

____ 21. What is the best way to approach a secluded structure at the end of a long, narrow driveway?

a. Drive forward as far as is safe and stop.
b. Run hoselines from the last known turnaround.
c. Back in from the last known turnaround.
d. Turn around at the structure before establishing attack lines.

____ 22. Which of the following is a correct procedure for attacking a secluded structure at the end of a long, narrow driveway?

a. Park on the roadway.
b. Park on the windward side of the structure.
c. Keep cab doors open.
d. Place the engine's air-conditioning system in recirculation mode.

____ 23. Which of the following statements about wildland apparatus is true?

a. IFSTA recommends that firefighters should not walk ahead of wildland apparatus as "spotters."
b. A short 2½-inch *(65 mm)* line should be deployed and charged for protection of the apparatus.
c. The shoulders of railroad roadbeds make good roadways for apparatus.
d. In a mobile attack, hoselines should be kept as short as possible.

____ 24. Which of the following is a valid safety guideline for using engines in the wildland environment?

a. Engine and crews should attack the front of a fire if it is spreading rapidly.
b. Apparatus should not be driven into unburned fuels higher than the bumper under any circumstances.
c. Headlights should be on whenever the engine is running.
d. Crews should work burned and unburned areas simultaneously.

____ 25. Personnel assigned to ____ commonly perform truck company functions when no aerial apparatus is on the scene.

a. Rescue/squad apparatus
b. Command vehicles
c. Breathing air supply apparatus
d. Emergency medical service vehicles

____ 26. Which of the following is the best position for a command vehicle?

a. Just inside the immediate danger zone at a corner of the building
b. Just inside the immediate danger zone at the front of the building
c. Outside the immediate danger zone at a corner of the building
d. Outside the immediate danger zone at the front of the building

____ 27. Flashing green lights commonly indicate ____.

a. Rescue/squad apparatus
b. Command vehicles
c. Breathing-air supply apparatus
d. Emergency medical service vehicles

____ 28. Which of the following statements about breathing air supply is true?

a. Breathing air compressors have a limited duration of use because they themselves must be refilled.
b. The positioning of mobile air supply apparatus is much the same as for command vehicles.
c. Apparatus using breathing air compressors may be parked downwind of the fire.
d. Breathing air compressors have filter sensors that prevent their use if the incoming air is contaminated.

B 29. NFPA ____ requires EMS vehicles to stand by at fire and hazardous materials incidents in case an emergency responder needs medical assistance.

a. 1004
b. 1500
c. 1005
d. 1400

____ 30. On incidents with no immediate EMS situations, EMS vehicles should be positioned in the ____ area.

a. Triage
b. Firefighter rehabilitation
c. Command
d. “Red”

____ 31. Level I staging requires later-arriving units to park ____.

a. At least one block away from the scene in their direction of travel
b. Less than one block away from the scene, opposite their direction of travel
c. At least one-half block away from the scene, opposite their direction of travel
d. At least two blocks away from the scene in their direction of travel

____ 32. Which of the following statements about operations on highways is true?

a. Responders should not use aerial apparatus or ground ladders to reach incidents on bridges.
b. Front wheels should be turned toward the firefighters working highway incidents.
c. All warning lights should be turned off at the scene to prevent blinding other drivers.
d. The use of warning lights and siren may create traffic conditions that actually slow the fire unit’s response.

____ 33. The contaminated area surrounding a hazardous material incident is the ____ zone.

a. Red zone
b. Limited-access zone
c. Yellow zone
d. Contamination reduction zone

____ 34. The warm zone is also called the ____ zone.

a. Exclusion zone
b. Restricted zone
c. Limited-access zone
d. Support zone

____ 35. Which of the following should the driver/operator attempt to do first when operating near a railroad?

a. Confirm from the rail company that train traffic has been halted on that set of tracks.
b. Run hose under the tracks.
c. Park the apparatus on the same side of the tracks as the incident.
d. Use aerial apparatus to run hose over the top of the area.

B_ 36. Which of the following is the best place to park an apparatus at an emergency medical incident?

a. Between smaller apparatus and oncoming traffic
b. Off the street
c. Between smaller apparatus and the emergency scene
d. Just past the emergency scene in the direction of traffic

Chapter 6

(A)

1 E
2 H
3 A
4 I
5 F
6 D
7 B
8 G

(B) 3

~~1 A~~ C 110
2 D
~~3 C~~ B 110
4 F
~~5 B~~ A 110

(C) 2

1 D
2 I
3 N
4 K
5 G
6 L
7 A
8 J
~~9 H~~ O 113
~~10 O~~ C 113
11 H
12 M
13 F
14 B

(D) 4

1 D
~~2 B~~ G 120
~~3 G~~ B
4 F
~~5 E~~ A 120
~~6 A~~ E 120

(E) 12

1 T
~~2 T~~ F
~~3 T~~
4 T
~~5 T~~ F
6 F
7 T
~~8 T~~ F
~~9 F~~ T
10 F
~~11 F~~ T 114
~~12 F~~ T 115
13 F
14 T
15 T
16 F
~~17 T~~ F
~~18 T~~ 121
19 F
20 T
~~21 F~~ T 122
~~22 T~~ F
~~23 F~~ T
~~24 F~~ T
25 T
26 T
~~27~~
~~28~~
~~29~~
~~30~~
~~31~~
~~32~~
~~33~~
~~34~~
~~35~~
~~36~~
~~37~~
~~38~~
~~39~~
~~40~~
~~41~~
~~42~~
~~43~~

(F)

1 C
2 C
3 B
4 A
5 D
6 ~~A~~ D
7 C
8 B
9 d
10 d
~~11 A~~ B 109
12 A
~~13 A~~ C 112
14 Not
15 B
~~16 G~~ A 114
~~17 B~~ D
18 C
~~19 D~~ B
~~20 C~~ A
~~21 C~~ D
22 C
~~23 A~~ D
~~24 D~~ C
25 B
~~26 C~~ D
~~27 B~~ A
28 B
29 B
~~30 D~~ C
31 D
~~32 C~~ B
33 C
~~34 D~~ C
35 C
36 A
~~37 A~~ C
38 D
39 B
40 B
~~41 C~~ d
~~42 A~~ d
~~43 A~~ C
44 A
~~45 B~~ C
~~46 D~~ B

Chapter 6 What Is Water and Where Does It Come From?

Matching

A. Match to their definitions terms associated with characteristics of water. Write the appropriate letters on the blanks.

____ 1. Compound formed when two parts of hydrogen combine with one part of oxygen

____ 2. Absorbing heat from a fire

____ 3. Excluding oxygen from a fire

____ 4. Ratio between the amount of heat needed to raise the temperature of a specified quantity of a material and the amount of heat needed to raise the temperature of an identical quantity of water the same number of degrees

____ 5. Quantity of heat a substance absorbs when it changes from a liquid to a vapor

____ 6. Temperature at which a liquid absorbs enough heat to change to vapor

____ 7. The tendency of a liquid to possess internal resistance to flow

____ 8. The density of liquids in relation to water

a. Smothering
b. Viscosity
c. Dissipating
d. Boiling point
e. Water
f. Latent heat of vaporization
g. Specific gravity
h. Cooling
i. Specific heat

B. Match to their definitions principles of pressure. Write the appropriate letters on the blanks. Definitions continued and terms repeated on the next page.

C 1. Pressure applied to a confined fluid from without is transmitted equally in all directions.

____ 2. The pressure of a liquid in an open vessel is proportional to its depth.

B 3. Fluid pressure at a point in a fluid at rest is the same intensity in all directions.

a. First principle
b. Second principle
c. Third principle
d. Fourth principle
e. Fifth principle
f. Sixth principle

____	4. The pressure of a liquid on the bottom of a vessel is independent of the shape of the vessel.	a. First principle
A	5. Fluid pressure is perpendicular to any surface on which it acts.	b. Second principle
		c. Third principle
		d. Fourth principle
		e. Fifth principle
		f. Sixth principle

C. **Match to their definitions terms related to water pressure and velocity. Write the appropriate letters on the blanks.**

____	1. Force per unit area	a. Residual pressure
____	2. Measure of weight	b. Water hammer
____	3. Any pressure less than atmospheric pressure	c. Altitude
____	4. Absolute zero pressure	d. Pressure
____	5. Height of a water supply above the discharge orifice	e. Velocity
____	6. Stored potential energy available to force water through pipe, fittings, fire hose, and adapters	f. Critical velocity
____	7. That part of total available pressure not used to overcome friction loss or gravity	g. Head
____	8. Forward velocity pressure at a discharge opening while water is flowing	h. Elevation pressure
O	9. Center line of pump or bottom of static water supply source above or below ground level	i. Force
C	10. Position of an object above or below sea level	j. Flow pressure
____	11. Losses or gains in pressure due to gravity	k. Perfect vacuum
____	12. That part of total pressure lost while forcing water through pipe, fittings, fire hose, and adapters	l. Static pressure
		m. Friction loss
		n. Vacuum
		o. Elevation

____ 13. Point at which friction becomes so great that the resistance agitates the entire stream

____ 14. Surge of energy in the opposite direction caused by suddenly stopping water moving through a hose or pipe

D. Match to their definitions terms related to water supply systems. Write the appropriate letters on the blanks.

____ 1. The part of the overall system that receives water from the pumping station and delivers it throughout the area served

____ 2. Hydrant that receives water from two or more directions

____ 3. Distribution system that provides circulating feed from several mains

____ 4. Large pipes that convey large quantities of water to various points for local distribution to smaller mains

____ 5. Network of intermediate-sized pipes that reinforce grid and aid concentration of fire flow at any point

____ 6. Smaller mains serving individual fire hydrants and blocks of consumers

a. Secondary feeders
b. Grid system
c. Dead-end hydrant
d. Distribution system
e. Distributors
f. Primary feeders
g. Circulating feed or looped line

True/False

E. Write *True* or *False* before each of the following statements. Correct those statements that are false.

________ 1. Water is by far the most commonly used extinguishing agent in today's fire service.

____

____

________ 2. In its vapor form, water is visible as steam.

____

____

_______ 3. For practical purposes, water is incompressible.

_______ 4. Water absorbs 8.5 times as much heat as does an equal amount of solid carbon dioxide.

F ___ 5. The speed with which water absorbs heat decreases in proportion to the amount of water surface exposed to the heat.

_______ 6. Water is an effective agent for extinguishing alcohol fires.

_______ 7. The readings of most pressure gauges are psi (or kPa) in addition to the existing atmospheric pressure.

F 8. The pressure in a water system before water flows from a hydrant is true static pressure.

T ___ 9. Altitude's effect on a fire stream's pressure is of little consequence up to about 2,000 feet.

_______ 10. The lower atmospheric pressure at high elevations reduces the amount of work needed to produce the pressure for an effective fire stream.

_____ 11. The movement of water molecules against each other causes friction loss.

_____ 12. The smaller the hose, the greater the velocity needed to deliver the same volume of water.

_____ 13. Hose brand does not affect friction loss.

_____ 14. Proper hose handling techniques alone are usually sufficient to minimize sharp bends or kinks in hoses.

_____ 15. Water hammer can damage a municipal water system.

_____ 16. Fire safety engineers should be considered the experts in water system problems.

_____ 17. Most communities use either a direct pumping water supply system or a gravity system.

_____ 18. Fire department personnel should note the condition and accessibility of valves and make any needed repairs during fire hydrant inspections.

______ 19. An OS&Y valve's stem is inside the yoke when the valve is open.

______ 20. Nonindicating valves in a water supply system are normally buried or installed in manholes.

T 21. Friction loss due to partially closed valves might not be noticeable during domestic flows of water.

F 22. Most pipe materials offer roughly the same resistance to water flow.

T 23. Private water supply systems usually receive their water from a municipal water supply system.

T 24. Almost all private water supply systems have separate piping for fire protection and domestic/industrial purposes.

______ 25. One advantage of separate piping for fire protection and domestic/industrial services is that interruptions to one system do not affect service to the other.

______ 26. Many significant fire losses can be traced to the failure private water supply systems that municipal fire departments were using to work incidents.

Multiple Choice

F. Write the letter of the best answer on the blank before each statement.

____ 1. For fire protection purposes, ordinary fresh water is generally considered to weigh ____.

a. 60 lb/ft^3 *(0.96 kg/L)*
b. 61.4 lb/ft^3 *(0.99 kg/L)*
c. 62.5 lb/ft^3 *(1 kg/L)*
d. 32 lb/ft^3 *(0.51 kg/L)*

____ 2. The measure of a substance's heat-absorbing capacity is its ____.

a. Specific gravity
b. Latent heat
c. Specific heat
d. Thermal potential

____ 3. A Btu is ____.

a. Equal to 1 joule
b. The amount of heat required to raise the temperature of 1 pound of water 1°F
c. A unit of work that has replaced the calorie in the SI heat measurement
d. Equal to 4.19 joules

____ 4. Water's latent heat of vaporization is approximately ____.

a. 970 Btu *(1 023 kJ)*
b. 1,000 Btu *(1 055 kJ)*
c. 750 Btu *(790 kJ)*
d. 1,079 Btu *(1 672 kJ)*

____ 5. At 212°F *(100°C)* water expands to approximately ____ times its original volume.

a. 19
b. 100
c. 350
d. 1,700

____ 6. Why is water's expansion when converted to steam helpful in fighting fires?

a. It absorbs more heat.
b. It reduces the need for ventilation.
c. It absorbs poisonous fumes.
d. It displaces other gases.

____ 7. Which of the following is ***not*** an observable result of the proper application of a water fire stream into a room?

a. Fire is extinguished.
b. Fire is reduced.
c. Visibility is reduced.
d. Temperature is reduced.

____ 8. Most flammable liquids have a specific gravity ____.

a. Equal to 0
b. Less than 1
c. Equal to 1
d. Greater than 1

____ 9. Which of the following is an advantage of water as an extinguishing agent?
- a. Low heat-absorbing capacity
- b. Low surface tension
- c. High opacity and reflectivity
- d. High availability

____ 10. Which of the following is a ***disadvantage*** of water as an extinguishing agent?
- a. Low surface tension
- b. Nonreactive with combustible metals
- c. Excessive opacity
- d. Electrical conductivity

____ 11. How much pressure does a one-square-inch column of water that is one foot tall exert at its base?
- a. 62.5 psi
- b. 0.434 psi
- c. 1 psi
- d. 2.304 psi

____ 12. At sea level the atmosphere exerts a pressure of ____.
- a. 14.7 psi *(101 kPa)*
- b. 15.6 psi *(107 kPa)*
- c. 16.6 psi *(114 kPa)*
- d. 17.7 psi *(121 kPa)*

____ 13. If feet of head equals 90, head pressure equals ____.
- a. 34.6 psi
- b. 80 psi
- c. 39 psi
- d. 230.9 psi

____ 14. If kPa equals 500, meters of head equals ____.
- a. 50
- b. 40
- c. 500
- d. 1 154.50

____ 15. Flow pressure of a water stream can be measured by using a(n) ____.
- a. Anemometer
- b. Pitot tube
- c. psi/kPa knife
- d. Sphygmomanometer

____ 16. The friction loss in old hose may be as much as ____ percent greater than in new hose.
- a. 50
- b. 60
- c. 65
- d. 75

____ 17. The roughness of the inner surface of pipe is commonly called the ____.
- a. Surface ratio
- b. Pressure loss equivalent
- c. Internal resistance factor
- d. Coefficient of friction

____ 18. According to the first principle of friction loss, if all other conditions are the same, friction loss ____ the length of the hose or pipe.
- a. Varies inversely as
- b. Varies approximately with the square of
- c. Varies directly with
- d. Is approximately the same as

____ 19. According to the second principle of friction loss, when hoses are the same size, the friction loss ____ the increase in the velocity of the flow.

a. Varies inversely as
b. Varies approximately with the square of
c. Varies directly with
d. Is approximately the same as

____ 20. According to the third principle of friction loss, for the same discharge, friction loss ____ the fifth power of the diameter of the hose.

a. Varies inversely as
b. Varies approximately with the square of
c. Varies directly with
d. Is approximately the same as

____ 21. The fourth principle of friction loss states that for a given flow velocity, friction loss ____, regardless of the pressure on the water.

a. Varies inversely as
b. Varies approximately with the square of
c. Varies directly with
d. Is approximately the same

____ 22. Because water is practically ____, the same volume of water supplied into a fire hose under pressure at one end will be discharged at the other end.

a. Nonviscous
b. Unstoppable
c. Incompressible
d. 100 percent saturated

____ 23. By using a nozzle to decrease the amount of flow, a firefighter ____ the speed of the water in the hose; consequently , there is ____ friction loss.

a. Increases/more
b. Increases/less
c. Decreases/more
d. Decreases/less

____ 24. What is the largest diameter hose than can realistically be used for hand lines?

a. 2 inches *(50 mm)*
b. 2½ inches *(65 mm)*
c. 3 inches *(77 mm)*
d. 3½ inches *(90 mm)*

____ 25. What is the best way to prevent water hammer?

a. Remove pressure relief devices from apparatus.
b. Open and close valves and nozzles slowly.
c. Use larger diameter hoses.
d. Reduce pressure in the water supply system.

____ 26. Which of the following is an example of ground water?

a. Reservoir
b. Cistern
c. River
d. Spring

____ 27. Which of the following systems uses one or more pumps to take up water from the primary source and force it into the distribution system?

a. Direct pumping
b. Gravity
c. Power siphon
d. Indirect gravity

____ 28. Which of the following systems uses a primary water source located at a higher elevation than the distribution system?

a. Direct pumping
b. Gravity
c. Power siphon
d. Indirect gravity

____ 29. In most combination systems, gravity flow is supplied by ____.

a. Hilltop ponds
b. Elevated storage tanks
c. Underground cisterns
d. Aqueducts

____ 30. Which of the following statements is true?

a. A good combination system does not involve duplicated equipment.
b. Fire departments should not use private industrial water supply systems.
c. Elevated storage tanks feed water back into the system during high-use times.
d. An efficient combination water supply system provides only enough water for domestic and fire fighting demands.

____ 31. What is the fire department's main concern regarding water treatment facilities?

a. Additives that affect water's fire fighting characteristics
b. Removal of materials that affect water's fire fighting characteristics
c. The combination of additives and removal of materials that affect water's fire fighting characteristics
d. Reduced volume and pressure of water available for fire fighting operations

____ 32. What is the recommended minimum size for fire hydrant supply mains?

a. 3 inches *(77 mm)*
b. 6 inches *(150 mm)*
c. 8 inches *(200 mm)*
d. 12 inches *(300 mm)*

____ 33. Fire hydrant supply mains should be gridded by cross-connecting mains an intervals of not more than ____.

a. 200 feet *(60 m)*
b. 300 feet *(90 m)*
c. 600 feet *(180 m)*
d. 750 feet *(225 m)*

____ 34. What is the recommended minimum size for fire hydrant supply mains in business and industrial districts?

a. 3 inches *(77 mm)*
b. 6 inches *(150 mm)*
c. 8 inches *(200 mm)*
d. 12 inches *(300 mm)*

____ 35. At a minimum, how often should a water distribution system's valves be operated?

a. Once a month
b. Once every six months
c. Once every year
d. Once every two years

____ 36. Which of the following valve types displays the words *open* and *shut?*

a. PIV
b. Gate
c. OS&Y
d. Butterfly

____ 37. What indicates whether a nonrising-stem gate valve is opened or closed?

a. Visual flow observation
b. Control wheel markings
c. Number of turns
d. Audible indicator

____ 38. Which of the following valve types operates by rotating the valve disk 90 degrees from fully open to fully closed?

a. PIV
b. Gate
c. OS&Y
d. Butterfly

____ 39. In a well designed water supply system, repairing a single line break will usually necessitate closing ___ hydrant(s), at most.

a. One
b. Two
c. Three
d. Four

____ 40. Which of the following is ***not*** a common material for underground pipe in water supply systems?

a. Ductile iron
b. Vitreous clay
c. Asbestos cement
d. Plastic

____ 41. Which of the following materials would be best for a water main in unstable soil?

a. Cast iron
b. Ductile iron
c. Plastic
d. Steel

____ 42. Average daily consumption is the average total amount of water used over a ____ period.

a. One-month
b. Three-month
c. Six-month
d. Twelve-month

____ 43. Maximum daily consumption is the maximum total amount of water that was used during any 24-hour interval within a ____ period.

a. One-year
b. Two-year
c. Three-year
d. Four-year

____ 44. The peak hourly rate normally varies from ____ times the normal hourly rate.

a. Two to four
b. Two to five
c. One and one-half to two
d. Five to ten

45. Which of the following is not a common purpose of a private water supply system?
    a. Providing water strictly for fire protection purposes
    b. Providing water for sanitary and fire protection purposes
    c. Providing water strictly for manufacturing purposes
    d. Providing water for fire protection and manufacturing processes

46. Why is nonpotable water in a private water system of special concern?
    a. Contaminated runoff
    b. Backflow into municipal system
    c. Damage to contents
    d. Sedimentation and encrustation

Identify

G. Identify the following abbreviations associated with water. Write the correct interpretation before each.

_______________ 1. psig

_______________ 2. psia

_______________ 3. PIV

_______________ 4. OS&Y

_______________ 5. ADC

_______________ 6. MDC

_______________ 7. PHC

Chapter 7

(A)

1 D
2 K
3 G
4 N
~~5 A~~ R 128
~~6 E~~ A 128
~~7 J~~ Q 129
~~8 F~~ J 129
9 F
10 B
11 L
12 C
13 I
14 O
~~15 Q~~ M 132
16 H
17 P

(B)

1 T
2 T
3 F
4 F
~~5 F~~ T 129
~~6 T~~ F 129
7 T
~~8 F~~ T 130
9 F
10 T
11 T
12 F
~~13 F~~ T 134
14 T
15 T

(C)

1 D
2 C
3 A
4 80 D
5 ~~42 D A~~ B 127
6 ~~Not~~
7 A
~~8 B~~ C 129
9 C
10 B
~~11 C~~ D 130
12 B
13 A ?
~~14 B~~ C 130
~~15 d~~ B 131
~~16 A~~ D 131
~~17 d~~ C 131
18 80 C
~~19 D~~ C 131
~~20 D~~ A 132
21 D
22 B
23 A
24 D
25 C
26 B
27 D
28 A
29 D
30 D
31 B
32 D
33 C
~~34 B~~ A 137
35 C Not

Chapter 7 Fire Hose Nozzles and Flow Rates

Matching

A. Match to their definitions terms associated with fire hose nozzles and flow rates. Write the appropriate letters on the blanks. Definitions continued and terms repeated on the next page.

____ 1. Stream of water or extinguishing agent after it leaves a fire hose and nozzle until it reaches the desired point.

____ 2. Fire stream produced from a fixed orifice, smoothbore nozzle.

____ 3. The line bounding a rounded surface

____ 4. A turning point or state of being turned

____ 5. To strike or dash about or against

____ 6. Stream produced by driving several jets of water together at an angle

____ 7. Nozzle designed to flow a specific amount of water at a specific nozzle discharge pressure on all stream patterns

____ 8. Nozzle that has a number of constant flow settings

____ 9. Variable flow nozzle that can change pattern and maintain the same nozzle pressure

____ 10. Nozzle designed to be placed on mobile attack lines

____ 11. Any fire stream too large to be controlled without mechanical aid

____ 12. Master stream device that allows stream direction and angle to be changed while water is being discharged

____ 13. Master stream device that is mounted on a fire apparatus deck and connected directly to the pump by permanent pipe

a. Impinging stream
b. Handline nozzle
c. Monitor
d. Fire stream
e. Periphery-deflected stream
f. Automatic nozzle
g. Periphery
h. Water curtain nozzle
i. Turret pipe
j. Manually adjustable nozzle
k. Solid stream
l. Master stream
m. Elevated master stream
n. Deflection
o. Deluge set
p. Nozzle reaction
q. Constant flow nozzle
r. Impinge

____ 14. Master stream device consisting of a short length of large diameter hose with a large nozzle or playpipe supported at the discharge end by a tripod

____ 15. Large capacity nozzle designed to be placed on the end of a fire department aerial apparatus elevating device

____ 16. Nozzle that produces a fan-shaped stream

____ 17. Counterforce created when water is discharged from a nozzle

a. Impinging stream
b. Handline nozzle
c. Monitor
d. Fire stream
e. Periphery-deflected stream
f. Automatic nozzle
g. Periphery
h. Water curtain nozzle
i. Turret pipe
j. Manually adjustable nozzle
k. Solid stream
l. Master stream
m. Elevated master stream
n. Deflection
o. Deluge set
p. Nozzle reaction
q. Constant flow nozzle
r. Impinge

True/False

B. Write *True* or *False* before each of the following statements. Correct those statements that are false.

________ 1. A fire stream is influenced by velocity, gravity, wind, and friction with the air.

____

____

________ 2. The condition of a stream when it leaves the nozzle is influenced by operating pressures, nozzle design, nozzle adjustment, and the condition of the nozzle orifice.

____

____

_____ 3. A wide-angle deflection produces a narrow-angle fog.

_____ 4. Impinging stream nozzles usually produce a narrow fog pattern.

T 5. Most constant flow nozzles use a periphery deflected stream.

F 6. Manually adjustable nozzles provide the indicated flow rate independent of pressure supplied by the driver/operator.

_____ 7. An automatic nozzle maintains approximately the same pressure and pattern even if the gallonage supplied to the nozzle changes.

T 8. A stream from an automatic nozzle can look good but may not be supplying sufficient water for extinguishment or protection.

_____ 9. High-pressure fog nozzles are not recommend for fighting wildland fires.

_____ 10. Master streams might be used when conditions are unsafe for an interior attack or when manpower is limited.

________ 11. To permit vertical and horizontal adjustments, master stream devices must route water through one or more sharp bends.

____

____

________ 12. Detachable ladder pipe streams can be moved horizontally.

____

____

T 13. Cellar nozzles may also be used to attack attic fires.

____

____

________ 14. The practical working limits for velocity of fire streams are within 60 to 120 feet per second.

____

____

________ 15. Fixed master stream devices with solid stream nozzles may be operated at pressures up to 100 psi *(700 kPa).*

____

____

Multiple Choice

C. Write the letter of the best answer on the blank before each statement.

____ 1. Which of the following is ***not*** a characteristic of a solid stream nozzle?

a. Compact

b. Little shower or spray

c. Can reach areas that other streams might not

d. Driven against an obstruction as it leaves the nozzle

____ 2. A short distance from the outlet, a solid stream nozzle becomes a cylindrical bore whose length is ____ times its diameter.

a. From one to two

b. Exactly one

c. From one to one and one-half

d. Exactly one and one-half

____ 3. On handlines, solid stream nozzles should be operated at ____.

a. 50 psi *(350 kPa)*
b. 60 psi *(420 kPa)*
c. 70 psi *(490 kPa)*
d. 80 psi *(560 kPa)*

____ 4. A solid master stream device should be operated at ____.

a. 50 psi *(350 kPa)*
b. 60 psi *(420 kPa)*
c. 70 psi *(490 kPa)*
d. 80 psi *(560 kPa)*

B 5. Water flow from a 1½-inch nozzle tip at 50 psi is approximately ____ gpm.

a. 420
b. 472
c. 315
d. 467

____ 6. Water flow from a 57 mm nozzle tip at 350 kPa is approximately ____ L/min.

a. 1 660
b. 4 020
c. 4 072
d. 6 067

____ 7. Which of the following does ***not*** directly affect the reach of a fog stream?

a. Water temperature
b. Stream width
c. Water droplet size
d. Amount of water flowing

C 8. Why are fog patterns usually easier to handle than solid or straight stream patterns?

a. Lower discharge pressure
b. Reduced volume
c. Counterbalanced reaction forces
d. Regulated velocity

____ 9. Most constant flow nozzles are intended to be operated at a nozzle pressure of ____.

a. 50 psi *(350 kPa)*
b. 75 psi *(535 kPa)*
c. 100 psi *(700 kPa)*
d. 150 psi *(1 050 kPa)*

____ 10. Most manually adjustable nozzles are intended to be operated at a nozzle pressure of ____.

a. 75 psi *(535 kPa)*
b. 100 psi *(700 kPa)*
c. 125 psi *(875 kPa)*
d. 150 psi *(1 050 kPa)*

D 11. Another name for an automatic nozzle is ____.

a. Fixed rate nozzle
b. Adjusting force nozzle
c. Self-regulating nozzle
d. Constant pressure nozzle

____ 12. Most automatic nozzles are designed for a ____ discharge pressure.

a. 75 psi *(535 kPa)*
b. 100 psi *(700 kPa)*
c. 125 psi *(875 kPa)*
d. 150 psi *(1 050 kPa)*

____ 13. What is the maximum volume if water supply to an automatic nozzle is inadequate?

a. That which can be achieved without the pump running away
b. One half the normal maximum volume
c. Just enough to allow the pump to cavitate
d. Seventy percent of normal maximum volume

____ 14. If pump discharge pressure increases, an automatic nozzle maintains constant nozzle pressure by ____.

a. Backflowing excess pressure
b. Decreasing its effective opening size
c. Enlarging its effective opening size
d. Modifying its fog pattern

____ 15. High-pressure fog nozzles operate at pressures up to ____.

a. 700 psi (4 900 kPa)
b. 800 psi (5 600 kPa)
c. 900 psi (6 300 kPa)
d. 1,000 psi (7 000 kPa)

____ 16. High-pressure fog nozzles generally flow approximately ____.

a. 5–10 gpm *(20 L/min to 40 L/min)*
b. 12–18 gpm *(48 L/min to 72 L/min)*
c. 10–20 gpm *(40 L/min to 80 L/min)*
d. 8–15 gpm *(32 L/min to 60 L/min)*

____ 17. What is generally the maximum amount of water that can safely flow through a handline nozzle?

a. 200 gpm *(800 L/min)*
b. 250 gpm *(1 000 L/min)*
c. 350 gpm *1 400 L/min)*
d. 450 gpm *(1 800 L/min)*

____ 18. Solid master streams are usually operated at ____.

a. 70 psi *(490 kPa)*
b. 75 psi *(525 kPa)*
c. 80 psi *(560 kPa)*
d. 85 psi *(495 kPa)*

____ 19. Master stream flows are usually ____ or greater.

a. 200 gpm *(800 L/min)*
b. 250 gpm *(1 000 L/min)*
c. 350 gpm *1 400 L/min)*
d. 450 gpm *(1 800 L/min)*

____ 20. The three basic types of monitors are ____.

a. Fixed, portable, and combination
b. Fixed, deluge set, and combination
c. Deluge set, portable, and combination
d. Fixed, portable, and deluge set

____ 21. Turret pipes are sometimes called ____.

a. Water cannon
b. Fire howitzers
c. Turret masters
d. Deck pipes

____ 22. A ____ is a master stream device used in conjunction with aerial apparatus ladders.

a. Step master
b. Ladder pipe
c. Aerial gun
d. Water cannon

____ 23. Which of the following is ***not*** used to power a remote-control ladder pipe?

a. Internal combustion
b. Hydraulics
c. Electricity
d. Pneumatics

____ 24. Thc primary difference between master stream devices used on aerial platforms and those used with prepiped aerial ladders is that aerial platform devices ____.

a. Can deliver higher pressures
b. Can not be operated by remote control
c. Require fewer personnel
d. Are more easily maneuvered

____ 25. Which of the following statements about special purpose nozzles is true?

a. NFPA 1901 requires each apparatus to carry at least one of every type of special purpose nozzle.
b. Broken streams produce smaller water droplets than fog streams.
c. Broken streams are created when water is forced through a series of small holes.
d. Broken streams have better reach than fog streams but less penetrating power.

____ 26. Which of the following is another name for a cellar nozzle?

a. Basement Betty
b. Distributor
c. Subfloor gun
d. Blind nozzle

____ 27. Which of the following statements about water curtain devices is true?

a. Water curtain nozzles are proven to be highly effective devices.
b. Water curtains absorb radiated heat.
c. Water curtains are more effective than flowing the same amount of water of the surface of a combustible material.
d. Water curtains absorb convected heat.

____ 28. Which of the following statements about piercing nozzles is true?
  a. The end of the piercing nozzle is driven with a sledge hammer.
  b. A periphery-deflected nozzle is built into the point of the piercing nozzle.
  c. Piercing nozzles should not be used to deliver aqueous film forming foam.
  d. The discharge end of a piercing nozzle is usually made of molybdenum.

____ 29. Which of the following statements about chimney nozzles is true?
  a. They should be lowered quickly down the chimney and then pulled slowly out.
  b. They deliver about 50 gpm *(200 L/min)*.
  c. Their sudden cooling effect damages the chimney flue liner.
  d. They produce a fine, misty fog cone.

____ 30. Practical working fire stream velocities are produced by nozzle pressures that range from ____.
  a. 50–150 psi *(350 kPa to 1 050 kPa)*
  b. 100–200 psi *(700 kPa to 1 400 kPa)*
  c. 20–50 psi *(140 kPa to 350 kPa)*
  d. 25–100 psi *(175 kPa to 700 kPa)*

____ 31. What is the maximum pressure at which solid stream handlines remain manageable?
  a. 50 psi *(350 kPa)*
  b. 65 psi *(455 kPa)*
  c. 75 psi *(525 kPa)*
  d. 80 psi *(560 kPa)*

____ 32. Solid stream nozzles used on aerial devices should be limited to a discharge pressure of ____.
  a. 50 psi *(350 kPa)*
  b. 65 psi *(455 kPa)*
  c. 75 psi *(525 kPa)*
  d. 80 psi *(560 kPa)*

____ 33. When must a driver/operator most commonly determine nozzle reaction?
  a. For a given hose layout on the fire scene
  b. Before practice or training sessions
  c. When determining configurations for preconnected attack lines
  d. During preincident planning for a specific structure

____ 34. What would nozzle reaction be (in pounds) for a 1¾-inch diameter solid stream nozzle delivering a stream at 65 psi?
  a. 312.5 pounds
  b. 245 pounds
  c. 178.6 pounds
  d. 374.7 pounds

____ 35. What would approximate nozzle reaction be (in newtons) for a fog stream nozzle delivering 400 L/min at 350 kPa?
  a. 2182 N
  b. 312.2 N
  c. 116.7 N
  d. 202.9 N

Chapter 8 Theoretical Pressure Calculations

True/False

A. Write *True* or *False* before each of the following statements. Correct those statements that are false.

_______ 1. The driver/operator's primary objective is to provide fire suppression crews with the water flow and pressure needed to achieve efficient fire control and/or extinguishment.

____

____

_______ 2. Calculations and field tests are equally accurate methods of determining friction loss.

____

____

_______ 3. The formula $2Q^2+Q$ provides the most accurate calculation of friction loss in today's fire hose.

____

_______ 4. Actual friction loss coefficients for hoses of the same diameter and length vary with manufacturers.

____

____

_______ 5. To calculate your own friction loss coefficients, you should always conduct tests on new hose.

____

____

_______ 6. Different types of hoses can be tested together for friction loss as long as they are the same diameter and have the same size couplings.

____

____

_________ 7. Test results are only as accurate as the devices used to take measurements.

____

____

_________ 8. Friction loss caused by handline nozzles is generally insignificant in an overall hose assembly.

____

____

_________ 9. Total pressure loss is the sum of friction loss in all hoses, appliances, and nozzles.

____

____

_________ 10. Because friction loss in hard piping is minimal, friction loss in standpipe is usually not significant enough to be included in friction loss calculations.

____

____

_________ 11. To supply multiple hoselines, the driver/operator should set the pump discharge pressure for the pump with the highest pressure demand and gate back the remaining hoselines at their discharge outlets.

____

____

Multiple Choice

B. Write the letter of the best answer on the blank before each statement.

____ 1. Which of the following is ***not*** a reason why driver/operators should be able to calculate theoretical friction loss?

a. It provides a better understanding of methods actually used on the fireground.
b. It is a skill they will use often on the fireground.
c. It allows them to predetermine discharge pressures for preconnected hoselines.
d. It serves as a tool for pre-incident planning.

____ 2. The primary determinant of friction loss is ____.
   a. Hose conditions
   b. Coupling conditions
   c. Volume of water flowing per minute
   d. Kinks

____ 3. Hills, gullies, aerial devices, or multistoried buildings create ____.
   a. Altitude pressure
   b. Platform differential
   c. Elevation pressure
   d. Grade coefficient

____ 4. Combined friction loss and pressure lost due to differences in height are referred to as ____.
   a. Negative gain
   b. Complete tension reduction
   c. Partial pressure
   d. Total pressure loss

____ 5. Which of the following is ***not*** a significant factor in calculating friction loss?
   a. Water temperature
   b. Hose length
   c. Hose diameter
   d. Major hose appliances

____ 6. Which of the following nozzle types should be used to determine the friction loss coefficients using a pitot tube?
   a. Smoothbore
   b. Fog
   c. Broken stream
   d. Any type

____ 7. Which of the following nozzle types should be used to determine the friction loss coefficients using a flowmeter?
   a. Smoothbore
   b. Fog
   c. Broken stream
   d. Any type

____ 8. ***(US)*** In what minimum increments should in-line gauges be calibrated for tests to determine friction loss coefficients?
   a. 1 psi
   b. 5 psi
   c. 10 psi
   d. 25 psi

____ 8. ***(Metric)*** In what minimum increments should in-line gauges be calibrated for tests to determine friction loss coefficients?
   a. 100 kPa
   b. 50 kPa
   c. 20 kPa
   d. 10 kPa

____ 9. Appliance friction loss is insignificant in cases where total flow through the appliance is less than ____.
   a. 250 gpm *(1 000 L/min)*
   b. 300 gpm *(1 200 L/min)*
   c. 350 gpm *(1 400 L/min)*
   d. 400 gpm *(1 600 L/min)*

____ 10. Generally it is safe to assume a friction loss of ____ or greater when water is flowing at an appliance's rated capacity.

a. 10 psi *(70 kPa)*
c. 20 psi *(140 kPa)*
b. 15 psi *(105 kPa)*
d. 25 psi *(175 kPa)*

____ 11. ***(US)*** Water exerts a pressure of ____ psi per foot of elevation.

a. 0.314
c. 0.447
b. 0.434
d. 2.301

____ 11. ***(Metric)*** Water exerts a pressure of ____ kPa per meter of elevation.

a. 5
c. 20
b. 10
d. 100

____ 12. Which of the following is a simple hose layout?

a. Equal length siamesed hoselines
b. Standpipe operation
c. Manifold hoselines of unequal length
d. Master stream

____ 13. ***(US)*** What is total pressure loss in a system if a pumper is directly supplying two lines that each have a friction loss of 36 psi and two that each have a friction loss of 47 psi?

a. 47 psi
c. 94 psi
b. 83 psi
d. 166 psi

____ 13. ***(Metric)*** What is total pressure loss in a system if a pumper is directly supplying two lines that each have a friction loss of 159.2 kPa and two that each have a friction loss of 273.3 kPa?

a. 273.3 kPa
c. 546.6 kPa
b. 432.5 kPa
d. 865.0 kPa

____ 14. Which of the following measurements should be used to calculate friction loss if the hoses supplying a master stream device are of different lengths?

a. The length of the longest hose
b. The length of the shortest hose
c. To total length of all hoses
d. The average length of all hoses

____ 15. How much pressure loss should be added to any calculations involving master stream appliances?

a. 10 psi *(70 kPa)*
b. 25 psi *(175 kPa)*
c. 30 psi *(210 kPa)*
d. 50 psi *(350 kPa)*

____ 16. If a pumper is discharging water at 150 psi *(1 050 kPa)* and the incoming pressure from the hydrant is 30 psi *(210 kPa)*, what is the net pump discharge pressure?

a. 180 psi *(1 260 kPa)*
b. 30 psi *(210 kPa)*
c. 120 psi *(840 kPa)*
d. 150 psi *(1 050 kPa)*

Calculate (US) *

C. Use the appropriate equations and tables from the manual to calculate values for the following situations. Round to the nearest tenth. Write your answers in the blanks.

*Slight differences in rounding may create variances in your answers and those listed in the Answers section of this study guide. If your answer is within 1 psi of the answer shown in the back of this study guide, you have likely solved the problem correctly.

1. If 250 gpm is flowing from a nozzle, what is the friction loss for 300 feet of 3-inch hose with 2½-inch couplings?

2. What is total elevation pressure gain for a hoseline operating at the bottom of a 79-foot hill?

3. What is total elevation pressure loss at the base of a hose connected to a building's standpipe system on the sixth floor and operating on the seventh floor?

4. What is total pressure loss when a pumper is supplying a 250-foot hoseline with 150 gpm, if the hoseline comprises 150 feet of 2½-inch hose reduced to 100 feet of 1¾-inch hose with 1½-inch couplings?

5. What is total pressure loss when a pumper is supplying a 200-foot length of 1½-inch hoseline with 125 gpm, if the hoseline is operating at the top of an 85-foot hill?

6. Calculate total friction loss for two 2½-inch hoselines, each 350 feet long with a 1¼-inch nozzle at a discharge pressure of 50 psi.

7. Calculate total friction loss for a wyed hose assembly if two 2-inch attack lines are wyed off with 300 feet of 3-inch hose with 3-inch couplings. Each 2-inch hose is 150 feet long and flowing 130 gpm.

8. What is the pressure loss in a hose assembly in which two 3-inch hoses with 2½-inch couplings, each 700 feet long, supply a siamese to which 250 feet of 2½-inch hose is attached. The 1½-inch tip has a nozzle pressure of 50 psi; it is 45 feet downhill from the siamese.

9. What is the total pressure loss when 100 feet of 2-inch hose is attached to a five-inch standpipe on a building's seventh floor and advanced to fight a fire on the eighth floor if the 1¼-inch tip has a discharge pressure of 50 psi?

10. Determine the total pressure loss due to friction for a hoseline assembly in which one 350-foot, 4-inch hose is supplying three attack lines attached to a water thief. The first attack line is 100 feet of 2-inch hose flowing 180 gpm; the second attack line is 150 feet of 1½-inch hose flowing 115 gpm; and the third attack line is 150 feet of 2-inch hose flowing 150 gpm.

11. What is total pressure loss in a hose assembly if three 4½-inch hoselines are supplying a 2,700 gpm master stream device? One hoseline is 500 feet, one is 600 feet, and one is 750 feet. Assume 900 gpm is flowing through each hoseline.

12. What pump discharge pressure is needed for 350 feet of 2-inch hose that is supplying a solid stream handline nozzle at 275 gpm?

13. How much pump discharge pressure is needed in a hoseline assembly when one 250-foot, 3-inch hoseline with 2½ -inch couplings is supplying three attack lines attached to a water thief if the first attack line is 150 feet of 1½-inch hose supplying 110 gpm, the second is 100 feet of 2-inch hose supplying 200 gpm, and the third is 200 feet of 2½-inch hose supplying 250 gpm? All nozzles are solid stream. The nozzles on lines 1 and 3 are on the top of a hill 20 feet above the pumper; the nozzle on line 2 is on the side of the hill, 15 feet above the pumper.

Calculate (Metric) *

C. Use the appropriate equations and tables from the manual to calculate values for the following situations. Round to the nearest kPa. Write your answers in the blanks.

*Slight differences in rounding may create variances in your answers and those listed in the Answers section of this study guide. If your answer is within 10 kPa of the answer shown in the back of this study guide, you have likely solved the problem correctly.

1. If 1 100 L/min is flowing from a nozzle, what is the friction loss for 200 meters of 77 mm hose with 65 mm couplings?

2. What is total elevation pressure gain for a hoseline operating at the bottom of a 39 meter hill?

3. What is total elevation pressure loss at the base of a hose connected to a building's standpipe system on the sixth floor and operating on the seventh floor?

4. What is total pressure loss when a pumper is supplying a 75-meter hoseline with 700 L/min, if the hoseline comprises 45 meters of 65 mm hose reduced to 30 meters of 45 mm hose?

5. What is total pressure loss when a pumper is supplying a 60-meter length of 38 mm hoseline with 600 L/min, if the hoseline is operating at the top of a 26-meter hill?

6. Calculate total friction loss for two 65 mm hoselines, each 120 meters long with a 25 mm nozzle operating at 350 kPa.

7. Calculate total friction loss for a wyed hose assembly if two 65 mm attack lines are wyed off with 90 meters of 77 mm hose with 77 mm couplings. Each 65 mm hose is 45 meters long and flowing 600 L/min.

8. What is the pressure loss in a hose assembly in which two 77 mm hoses with 65 mm couplings, each 225 meters long, supply a siamese to which 90 meters of 65 mm hose is attached. The 38 mm tip has a nozzle pressure of 350 kPa; it is 15 meters downhill from the siamese.

9. What is the total pressure loss when 30 meters of 65 mm hose is attached to a 125 mm standpipe on a building's seventh floor and advanced to fight a fire on the eighth floor if the 32 mm tip has a discharge pressure of 350 kPa?

10. Determine the total pressure loss due to friction for a hoseline assembly in which one 15-meter, 100 mm hose is supplying three attack lines attached to a water thief. The first attack line is 30 meters of 50 mm hose flowing 700 L/min; the second attack line is 45 meters of 38 mm hose flowing 450 L/min; and the third attack line is 45 meters of 50 mm hose flowing 600 L/min.

11. What is total pressure loss in a hose assembly if three 115 mm hoselines are supplying a 10 500 L/min master stream device? One hoseline is 150 meters, one is 180 meters, and one is 235 meters. Assume 3 500 L/min is flowing through each hoseline.

12. What pump discharge pressure is needed for 100 meters of 50 mm hose that is supplying a solid stream handline nozzle at 1 200 L/min?

13. How much pump discharge pressure is needed in a hoseline assembly when one 90 meter, 77 mm hoseline with 65 mm couplings is supplying three attack lines attached to a water thief if the first attack line is 45 meters of 38 mm hose supplying 400 L/min, the second is 30 meters of 50 mm hose supplying 800 L/min, and the third is 60 meters of 65 mm hose supplying 1 000 L/min? All nozzles are solid stream. The nozzles on lines 1 and 3 are on the top of a hill 10 meters above the pumper; the nozzle on line 2 is on the side of the hill, 7 meters above the pumper.

Identify

D. Identify the following abbreviations associated with pressure calculations. Write the correct interpretation before each.

_______________ 1. TPL

_______________ 2. PDP

Chapter 9 Fireground Hydraulic Calculations

Matching

A. Match to their definitions terms associated with fireground hydraulic calculations. Write the appropriate letters on the blanks.

____ 1. Measures the amount of water going through a discharge in gallons (liters) per minute

____ 2. Enables the driver/operator to determine pump discharge pressure needed to supply a hose layout without performing mental calculations

____ 3. Lists required pump discharge pressures for a specific jurisdiction's hose lays and assemblies

a. Pump chart

b. Flowmeter

c. Pressure converter

d. Hydraulic calculator

True/False

B. Write *True* or *False* before each of the following statements. Correct those statements that are false.

________ 1. The ultimate purpose of all fireground hydraulic calculations is to discharge an appropriate amount of water from the pumper.

____

____

________ 2. Flowmeters enable the driver/operator to pump the correct volume of water to automatic nozzles without having to know the length of hoseline, the amount of friction loss, or the elevation pressure.

____

____

________ 3. A central flowmeter monitoring device may indicate the total amount of water that has flowed through the pump for the duration of an incident.

____

____

Chapter 8

(A)

1 T
2 F
3 F
4 T
5 T
6 F
7 T
8 F
9 F
10 F
11 T

(B)

1 D
2 D
3 C
4 D ?
5 A
6 A
7 B
8 C
9 B
10 A
11 D ?
12 B
13 D ?
14 A
15 B
16 C

Ⓐ Chapter 9

1 B
2 D
3 C

Ⓑ

1 T
2 F
3 T
4 T
5 F
6 T
7 F

Ⓒ

1 C
2 A
3 B
4 B
5 D
6 D
7 B
8 A
9 A
10 C
11 C
12 B
13
14
15
16
17
18
19
20
21
22
23
24
25

_____ 4. Hose and nozzle manufacturers usually provide manual and/or mechanical hydraulic calculators to fire departments.

____

____

_____ 5. Programmable calculators that can be preprogrammed for fireground calculations are no longer manufactured.

____

____

_____ 6. The amount of friction loss calculated using the condensed "Q" formula for 3-inch hose will be 10 percent lower than if the same situation is calculated using $FL = CQ^2L$.

____

____

_____ 7. The gpm flowing method of determining friction loss does not apply to 2½-inch hose with flows below 160 gpm.

____

____

Multiple Choice

C. Write the letter of the best answer on the blank before each statement.

____ 1. NFPA 1901 allows flowmeters to be used instead of pressure gauges on all discharges from ____ in diameter.

a. 1 to 2½ inches *(25 to 65 mm)*
b. 1 to 3 inches *(25 to 77 mm)*
c. 1½ to 3 inches *(38 to 77 mm)*
d. 1½ to 4 inches *(38 to 100 mm)*

____ 2. Flowmeters must provide a readout in increments no larger than ____.

a. 10 gpm *(38 L/min)*
b. 25 gpm *(85 L/min)*
c. 50 gpm *(190 L/min)*
d. 100 gpm *(380 L/min)*

____ 3. The two types of flowmeters used in fire apparatus are ____.

a. Paddle wheel and windmill
b. Spring probe and paddle wheel
c. Windmill and flextube
d. Spring probe and flextube

____ 4. Properly calibrated flowmeters should be accurate to ±____ percent.

a. 1
b. 3
c. 5
d. 10

____ 5. Each discharge equipped with a flowmeter has a digital readout display mounted within ____ of its valve control.

a. 2 inches *(50 mm)*
b. 3 inches *(75 mm)*
c. 4 inches *(100 mm)*
d. 6 inches *(150 mm)*

____ 6. If water volume at the nozzle suddenly diminishes with no reduction in the flowmeter reading, you can assume that ____.

a. The flowmeter is not functioning properly.
b. The hose is kinked.
c. A midline valve is partially closed.
d. The hose has burst.

____ 7. To feed a relay pumping supply line without knowing the amount of water flowing from the pumper receiving the water, increase engine speed until ____.

a. The rpm will no longer increase.
b. The flowmeter reading drops slightly.
c. The flowmeter reading no longer increases.
d. The rpm equal the flow meter reading.

____ 8. To feed a relay pumping supply line without knowing the amount of water flowing from the pumper receiving the water, do not allow incoming pressure to drop much below ____.

a. 10 psi *(70 kPa)*
b. 20 psi *(140 kPa)*
c. 25 psi *(175 kPa)*
d. 50 psi *(350 kPa)*

____ 9. What is the best method to ensure that nozzle pressures are correct when automatic nozzles are used in multiple hoseline standpipe operations?

a. Communication
b. Flowmeter
c. Master intake pressure gauge
d. Hydraulic calculator

____ 10. Using the pump chart in Table 9.1 (page 196), determine the pump discharge pressure needed to supply 150 gpm to a 300-foot 2½-inch fog line.

a. 115 psi
b. 116 psi
c. 118 psi
d. 130 psi

____ 11. Using the pump chart in Table 9.1 (page 196), determine the pump discharge pressure needed to supply 375 gpm to a 200-foot hose lay with one 1½-inch line and one 2½-inch line attached to a standpipe system on a building's tenth floor to fight a fire on the eleventh floor.

a. 238 psi
b. 243 psi
c. 288 psi
d. 293 psi

____ 12. Using the pump chart in Table 9.2a (page 197), determine the pump discharge pressure needed to supply 750 gpm to a 700-foot hose relay with two 3-inch lines if the pumper receiving the water is at the bottom of a hill 40 feet lower than the pumper supplying the water.

a. 99 psi
b. 79 psi
c. 88 psi
d. 119 psi

____ 13. Using the pump chart in Table 9.2b (page 198), determine the pump discharge pressure needed to supply 1 000 L/min to a 65 mm fog nozzle on a wyed line in a 180 m hose lay.

a. 1 175
b. 1 270
c. 1 360
d. 1 365

____ 14. The hand method is useful for determining friction loss in ____ hose.

a. 1½-inch
b. 2-inch
c. 2½-inch
d. 3-inch

____ 15. Use the hand method to estimate friction loss in 400 feet of 2½-inch hose at 300 gpm.

a. 18 psi
b. 36 psi
c. 72 psi
d. 1800 psi

____ 16. Use the hand method to estimate friction loss in 200 feet of 1¾-inch hose at 175 gpm.

a. 18 psi
b. 36 psi
c. 72 psi
d. 1800 psi

____ 17. In the condensed "Q" formula, Q = ____

a. psi
b. gpm
c. gpm divided by 10
d. Hundreds of gpm

____ 18. Which of the following is the condensed "Q" formula for 3-inch hose?

a. FL per 100 feet = Q^2
b. FL per 100 feet = $Q^2/5$
c. FL per 100 feet = $Q^2/10$
d. FL per 100 feet = $Q^2/15$

____ 19. Which of the following is the condensed "Q" formula for 4-inch hose?

a. FL per 100 feet = Q^2
b. FL per 100 feet = $Q^2/5$
c. FL per 100 feet = $Q^2/10$
d. FL per 100 feet = $Q^2/15$

____ 20. Use the condensed "Q" formula to determine friction loss per 100 feet flowing 800 gpm in 500 feet of 5-inch hose. Round your answer to the nearest whole number.

a. 20 psi
b. 12 psi
c. 13 psi
d. 43 psi

____ 21. The gpm flowing method of determining friction loss for 2½-inch hose can be expressed as the formula ____ , where FL = friction loss per 100 feet of hose.

a. FL = gpm/10
b. FL = gpm × 100
c. FL = gpm/100
d. FL = (gpm/10) × 10

____ 22. Use the gpm flowing method to determine friction loss in 500 feet of 2½-inch hose flowing 270 gpm.

a. 17 psi
b. 27 psi
c. 85 psi
d. 125 psi

____ 23. At flow rates from 50 to 75 gpm, friction loss in 1½-inch hose equals the friction loss of ____ times the flow through 2½-inch hose.

a. 2
b. 3
c. 4
d. 5

____ 24. At flow rates from 80 to 100 gpm, friction loss in 1½-inch hose increases ____ psi for every 20 gpm increase in flow through 2½-inch hose.

a. 2
b. 3
c. 4
d. 5

____ 25. Using the gpm flowing method, what is the friction loss in 200 feet of 1½-inch hose flowing 95 gpm.

a. 29 psi
b. 32 psi
c. 58 psi
d. 64 psi

Chapter 10 Fire Pump Theory

Matching

A. Match to their definitions terms associated with fire pumps. Write the appropriate letters on the blanks. Definitions continued and terms repeated on the next page.

____ 1. Forces all water and air out of the pump body with each operating cycle

____ 2. Major pump on the modern fire department pumper

____ 3. Has four valves to produce a constant stream of water

____ 4. Consists of two gears that rotate in a tightly meshed pattern inside a watertight case

____ 5. Constructed with movable elements that automatically compensate for wear

____ 6. Imparts velocity to water and converts it to pressure within the pump itself

____ 7. Transmits energy to water in the form of velocity

____ 8. Converts velocity to pressure

____ 9. Cancels lateral thrust by taking in water from both sides

____ 10. Has two impellers mounted within a single housing

____ 11. Seals the space where the impeller shaft passes through the pump casing

____ 12. Pumping water while the apparatus is moving

____ 13. Allows the tank to be filled without making additional connections when the pump is being supplied from an external source

a. Casing
b. Rotary gear pump
c. Single-action piston pump
d. Tank fill line
e. Marine type
f. Packing ring
g. Pressure governor
h. Positive displacement pump
i. Impeller
j. Double-acting piston pump
k. Voltmeter
l. Two-stage centrifugal pump
m. Push-pull
n. Tachometer
o. Double suction impeller
p. Centrifugal pump
q. Parallel pump
r. Pump drain valve
s. Rotary vane pump
t. Pump and roll

____ 14. Valve actuator that uses a sliding gear-tooth rack that engages a sector gear connected to the valve stem

____ 15. Allows the operator to bleed pressure off the hoseline without leaving the apparatus to open the nozzle

____ 16. Regulates the engine's power output to match pump discharge requirements

____ 17. Records engine speed in revolutions per minute

____ 18. Indicates battery condition and alternator output

____ 19. Auxiliary cooling device that contains a number of small tubes surrounded by a water jacket

a. Casing
b. Rotary gear pump
c. Single-action piston pump
d. Tank fill line
e. Marine type
f. Packing ring
g. Pressure governor
h. Positive displacement pump
i. Impeller
j. Double-acting piston pump
k. Voltmeter
l. Two-stage centrifugal pump
m. Push-pull
n. Tachometer
o. Double suction impeller
p. Centrifugal pump
q. Parallel pump
r. Pump drain valve
s. Rotary vane pump
t. Pump and roll

True/False

B. Write *True* or *False* before each of the following statements. Correct those statements that are false.

________ 1. Most modern municipal water systems can maintain adequate pressure in the hydrant system for effective fire fighting without the need for additional pumps.

____

____

_________ 2. Modern apparatus still have some form of positive pressure pump connected to the main fire pump.

____

____

_________ 3. The cylinder's size usually determines a piston pump's capacity.

____

____

_________ 4. Piston pumps are practically impervious to wear.

____

____

_________ 5. The amount of water that a rotary gear pump can pump is determined by the size of the pockets in the gears and the speed of rotation.

____

____

_________ 6. Rotary gear pumps are very susceptible to damage from normal wear.

____

____

_________ 7. In a centrifugal pump, the larger the impeller's eye the lower the flow capacity.

____

____

_________ 8. Modern centrifugal pumps are self-priming.

____

____

_________ 9. If there is any question about the proper operation of the transfer valve, it is better to be in series mode than in parallel.

____

____

_________ 10. Power-operated transfer valves in centrifugal pumps may be safely operated at pressures as high as 200 psi *(1 400 kPa)*.

____

____

_________ 11. When a two-stage centrifugal pump is in pressure mode its clapper valves should be closed.

____

____

_________ 12. One approach to providing high discharge pressures is to supply a single-stage, high-pressure centrifugal pump mounted outboard on a conventional two-stage pump with a separate drive system.

____

____

_________ 13. If an impeller's hub becomes worn, the impeller must be replaced.

____

____

_________ 14. A driver/operator can check pump temperature by feeling the intake piping.

____

____

_________ 15. In general, when a centrifugal pump is operating under pressure, water may drip from the packing gland but should not run in a steady stream.

____

____

_________ 16. If a power take-off drive pump is used for pump and roll operations, the vehicle should be driven by the pressure gauge instead of the speedometer while hoselines are in operation.

____

____

_______ 17. A lock to prevent the road transmission from being engaged while a front-mounted pump is operating is a necessity on an automatic transmission and a recommended option on a manual transmission.

____

____

_______ 18. Most fire pumps can operate over a range of capacities limited only by engine horsepower and pump size.

____

____

_______ 19. Some automatic transmissions can momentarily engage in the lowest and highest forward ranges while internal clutches are being applied and released.

____

____

_______ 20. The piping system on a fire apparatus must be able to withstand a hydrostatic test of 500 psi *(3 450 kPa)* before being placed into service.

____

____

_______ 21. Midship pumps typically have a large intake connection on either side of the apparatus.

____

____

_______ 22. According to NFPA 1901, enough 2½-inch *(65 mm)* or larger discharge outlets must be provided to flow twice the rated capacity of the fire pump.

____

____

_______ 23. Discharges to which 1½, 1¾, and 2-inch *(38 mm, 45 mm, and 50 mm)* handlines are attached must be supplied by at least 2½-inch *(65 mm)* piping.

____

____

_________ 24. The tank fill line is the most effective way to circulate water through the pump to prevent overheating when no lines are flowing.

____

____

_________ 25. The quarter-turn valve handle uses a simpler mechanical linkage than a push-pull handle.

____

____

_________ 26. Simply opening the drain valve allows all water to drain from booster hoselines during freezing weather.

____

____

_________ 27. A discharge pressure relief valve is an integral part of all fire pumps that are not equipped with a pressure governor.

____

____

_________ 28. Intake pressure relief valves may be either built into a pump by the manufacturer or added on by the department.

____

____

_________ 29. Electronic governors provide satisfactory service but are not as fast or accurate as mechanical governors.

____

____

_________ 30. One advantage of driving a primer with an electric motor is that it can be operated effectively regardless of engine speed.

____

____

_________ 31. According to NFPA 1901, the primer control is a recommended option on the pump operator's panel.

____

____

_________ 32. An engine that operates too cool is inefficient.

____

____

_________ 33. The pump overheat indicator must provide an audible warning when the pump overheats.

____

____

_________ 34. The pumping engine oil pressure indicator measures the oil level in the crankcase.

____

____

Multiple Choice

C. Write the letter of the best answer on the blank before each statement.

____ 1. Positive placement pumps are used as priming pumps because ____.
- a. They are more efficient than centrifugal pumps.
- b. They exclude air.
- c. They can pump air.
- d. They can pump small amounts of water.

____ 2. Removing the trapped air from a positive displacement pump allows ____ to force water into the pump casing.
- a. Water pressure
- b. Gravity
- c. Centrifugal force
- d. Atmospheric pressure

____ 3. Which of the following pump types relies on reciprocating (back and forth) action to develop pressure?
- a. Piston
- b. Centrifugal
- c. Rotary
- d. Interpolating

____ 4. On its ____ stroke, a single-action piston pump discharges air or water.
- a. Return
- b. Clockwise
- c. Counterclockwise
- d. Forward

____ 5. On its ____ stroke, a single-action piston pump draws air or water into the cylinder.
- a. Return
- b. Clockwise
- c. Counterclockwise
- d. Forward

____ 6. With a double-acting pump, output is ____.

a. Constant
b. A series of pressure surges with one period of no flow
c. A series of pressure surges with two periods of no flow
d. A series of pressure surges with four periods of no flow

____ 7. Which of the following is an advantage of a multicylinder piston pump?

a. All cylinders are always engaged.
b. Provides a uniform discharge.
c. Piston speed is not a concern.
d. Intermittent discharge enhances fog streams.

____ 8. The most common application for high-pressure piston pumps that are still in service is ____.

a. High-rise fire fighting
b. Wildland fire fighting
c. Combustible liquid spill containment
d. Combustible liquid fire fighting

____ 9. Which of the following fire pumps has the simplest design?

a. Piston
b. Centrifugal
c. Rotary
d. Gravity

____ 10. Where is pressure greatest in a rotary vane pump?

a. At the discharge opening
b. Midway between the intake opening and the discharge opening
c. At the intake opening
d. Equal through the pump housing

____ 11. The height to which pressure rises in a centrifugal pump is determined by ____.

a. Size of the impeller
b. Size of the casing
c. Speed of rotation
d. Speed of reciprocation

____ 12. An impeller's speed in a centrifugal pump usually ranges from ____.

a. 1,000 to 2,000 rpm
b. 1,000 to 4,000 rpm
c. 2,000 to 5,000 rpm
d. 2,000 to 4,000 rpm

____ 13. The area in a centrifugal pump where water volume gradually increases is the ____.

a. Vane
b. Volute
c. Eye
d. Stripping edge

____ 14. Which of the following is ***not*** a main factor in a centrifugal pump's discharge pressure?
a. Amount of water discharged
b. Speed at which the impeller turns
c. Pressure of water when it enters the pump
d. Elevation pressure in discharge hose

____ 15. The increase in a centrifugal pump's discharge pressure is approximately equal to ____ the change in impeller speed.
a. The square of
b. One half of
c. Twice
d. The square root of

____ 16. Single stage centrifugal pumps with a single intake impeller and a simple casing can provide a maximum of ____.
a. 1,000 gpm *(4 000 L/min)*
b. 1,500 gpm *(6 000 L/min)*
c. 2,000 gpm *(8 000 L/min)*
d. 2,500 gpm *(10 000 L/min)*

____ 17. A two-stage centrifugal fire pump can deliver its greatest volume when it is in the ____ position.
a. Series
b. Proportional
c. Parallel
d. Sequencing

____ 18. A two-stage centrifugal fire pump can deliver its highest pressure when it is in the ____ position.
a. Series
b. Proportional
c. Parallel
d. Sequencing

____ 19. Most modern two-stage centrifugal pumps can remain in the pressure position until it is necessary to flow more than ____ the pump's rated capacity.
a. One third
b. One half
c. Two thirds
d. Three fourths

____ 20. The maximum net pressure at which a two-stage centrifugal pump's transfer valve should be operated is usually ____.
a. 50 psi *(350 kPa)*
b. 75 psi *(535 kPa)*
c. 100 psi *(700 kPa)*
d. 150 psi *(1 050 kPa)*

____ 21. The impeller hub usually must fit in the pump casing within a tolerance of ____ or less.
a. 0.001 inches *(0.025 mm)*
b. 0.0025 inches *(0.0625 mm)*
c. 0.01 inches *(0.25 mm)*
d. 0.025 inches *(0.625 mm)*

____ 22. What is the first indication that wear is becoming a problem in a pump.
a. Leaking from the pump housing during pump tests
b. Noise that decreases as pump speed increases
c. Vibrations that increase as pump speed increases
d. Increased engine rpm becomes necessary to pump the rated capacity

____ 23. If all discharges in a centrifugal pump are shut off, the energy being supplied to the impeller ____.

a. Returns back to the water supply
b. Usually causes the pump to explode
c. Displaces the wear rings
d. Dissipates in the form of heat

____ 24. A(n) ____ provides cooling and lubrication in a packing ring.

a. Lantern ring
b. Packing gland
c. Oil port
d. Grease fitting

____ 25. When does a centrifugal pump's packing receive the water it needs for lubrication?

a. When the pump is full and operating under pressure
b. When the pump is full but stationary
c. At all times
d. When the pump being primed

____ 26. After a period of dryness, when should a centrifugal pump's packing be adjusted?

a. Before operation
b. When the pump is operating but before the packing seals
c. When the pump is operating and the packing has sealed
d. After the pump has been used and shut down

____ 27. Which of the following is an advantage of mechanical seals over packing rings?

a. Mechanical seals do not have to be lubricated.
b. Mechanical seals do not have to be adjusted.
c. Packing seals are more susceptible to freezing.
d. Packing seals are more difficult to replace.

____ 28. Which of the following is the most important consideration when designing a pumper's pump mounting arrangements and drive systems?

a. Cost
b. Tradition
c. Use
d. Space required

____ 29. Which of the following is ***not*** a common application for an auxiliary engine-driven pump?

a. Structural pumper
b. Wildland fire apparatus
c. Airport rescue and fire fighting vehicles
d. Mobile water supply apparatus

____ 30. The capacity of auxiliary engine-driven pumps on wildland or mobile water supply apparatus is generally ____ or less.

a. 200 gpm *(800 L/min)*
b. 300 gpm *(1 200 L/min)*
c. 400 gpm *(1 600 L/min)*
d. 500 gpm *(2 000 L/min)*

____ 31. Which of the following is the ***least*** common use for a power take-off drive pump?

a. Initial attack apparatus
b. Structural pumper
c. Wildland pumper
d. Mobile water supply apparatus

____ 32. Which of the following statements about power take-off drive pumps is true?

a. The PTO shaft speed depends upon which road gear is in use.
b. The PTO shaft is independent of the clutch.
c. Skid plates do not provide acceptable protection if the pump extends too far below the chassis.
d. The pressure being developed changes when the driver changes the vehicle speed.

____ 33. The conventional power take-off unit limits pump capacity to about ____.

a. 250 gpm *(1 000 L/min)*
b. 500 gpm *(2 000 L/min)*
c. 750 gpm *(3 000 L/min)*
d. 1,000 gpm *(4 000 L/min)*

____ 34. The impeller of a front-mount pump turns at a speed that is ____ the engine's.

a. Independent of
b. Slower than
c. The same as
d. Faster than

____ 35. A front-mount pump is connected to ____.

a. The vehicle's transmission
b. The engine's camshaft
c. The engine's crankshaft
d. An auxiliary engine

____ 36. Which of the following statements about front-mount pumps is true?

a. Maximum capacity can go as high as 1,250 gpm *(5 000 L/min)* without major modifications.
b. Their susceptibility to freezing makes them unsuited to cold climates.
c. They cannot be used in pump and roll operations.
d. Properly mounted, they can withstand accidents that cause considerable damage to the apparatus.

____ 37. Where is the fire pump mounted on most fire department pumpers?

a. At the front of the apparatus
b. Midship behind the engine and in front of the transmission
c. Midship behind the engine and the transmission
d. At the rear of the apparatus

____ 38. Which of the following statements about midship transfer drive pumps is true?

a. If an apparatus is designed so that the speedometer does not go above zero when the pump is engaged, there is no practical way to tell visually if the transmission is in the correct gear.
b. The normal arrangement is for the transfer case to be controlled from outside the cab at the pump.
c. With the engine idling and the pump engaged, most speedometers read between 5 and 10 mph *(8 km/h and 16 km/h).*
d. To help with a smooth shift in and out of pump, the transmission should be running from 1½ to 2 times normal idle speed.

____ 39. Which of the following statements about rear-mount pumps is true?

a. They interfere with even weight distribution.
b. They allow more compartment space for tools and equipment than do midship pumps.
c. They help to protect the driver/operator from oncoming traffic.
d. They cannot be driven by a power take-off.

____ 40. The majority of all fires are initially fought with water from ____.

a. The fire apparatus tank
b. Portable extinguishers
c. A static source
d. A hydrant

____ 41. NFPA 1901 states that piping should be sized so that pumpers with a capacity of 500 gpm *(1 900 L/min)* or less can flow ____ from their booster tank.

a. 100 gpm *(380 L/min)*
b. 150 gpm *(570 L/min*
c. 200 gpm *(760 L/min))*
d. 250 gpm *(950 L/min)*

____ 42. NFPA 1901 states that piping should be sized so that pumpers with a capacity greater than 500 gpm *(1 900 L/min)* can flow at least ____ from their booster tank.

a. 250 gpm *(950 L/min)*
b. 500 gpm *(1 900 L/min)*
c. 750 gpm *(2 850L/min)*
d. 1,000 gpm *(3 800 L/min)*

____ 43. Which of the following statements about pump piping is true?

a. The pump can be primed even if the tank-to-pump valve leaks.
b. The pump must be primed when drafting from a static water supply source.
c. The intake line from the tank must enter the pump below the eye of the impeller.
d. The piping is typically round where it enters the pump.

____ 44. NFPA 1901 requires ____ suction line(s) for a pumper with a rated capacity of 1,750 gpm *(6 624 L/min)*.

a. One, 6-inch *(125-mm)*
b. One, 6-inch *(150-mm)*
c. Two, 6-inch *(150-mm)*
d. Three, 8-inch *(150-mm)*

____ 45. If 3-inch pipe is used and care is given to the fittings, it may be possible to flow as much as ____ through an additional intake line.

a. 300 gpm *(1 200 L/min)*
b. 450 gpm *(1 800 L/min)*
c. 500 gpm *(2 000 L/min)*
d. 650 gpm *(2 600 L/min)*

____ 46. NFPA 1901 specifies a flow rate of ____ for a 3½-inch *(89 mm)* diameter pump outlet.

a. 500 gpm *(1 900 L/min)*
b. 600 gpm *(2 280 L/min)*
c. 700 gpm *(2 660 L/min)*
d. 800 gpm *(3 040 L/min)*

____ 47. When multiple attack lines require different pressures, pump discharge pressure should be set at the ____.

a. Lowest required pressure
b. Average required pressure
c. Total required pressure
d. Highest required pressure

____ 48. NFPA 1901 requires apparatus with 1,000 gallon *(3 875 L)* or larger tanks to have a least a ____ tank fill line.

a. 1-inch *(25-mm)*
b. 1½-inch *(38-mm)*
c. 2-inch *(51-mm)*
d. 2½-inch *(65-mm)*

____ 49. If a booster line cooling valve is used to cool the pump when no lines are flowing, flow is limited to approximately ____.

a. 5 to 10 gpm *(20 L/min to 40 L/min)*
b. 10 to 15 gpm *(40 L/min to 60 L/min)*
c. 10 to 20 gpm *(40 L/min to 80 L/min)*
d. 15 to 25 gpm *(60 L/min to 100 L/min)*

____ 50. The most common type of valve in pump intake and discharge lines is the ____ type.

a. Gate
b. Butterfly
c. Plunger
d. Ball

____ 51. Gate valves on fire apparatus are most commonly operated by a ____.

a. Push-pull handle
b. Hand wheel
c. Quarter-turn handle
d. Scissors handle

____ 52. What equipment helps minimize the risk of water hammer by preventing movement from the fully opened to the fully closed position (or vice versa) in less than three seconds?

a. Ball speed regulator
b. Slow acting valve control
c. Slow ball
d. Retard valve

____ 53. Which of the following is ***not*** a purpose of a pump drain?
a. Provides auxiliary intake connection
b. Allows smooth transition of supply flow from tank to supply line
c. Bleeds off the discharge hose
d. Avoids frozen pipes

____ 54. Why should the drain valve not be opened when the pump is operating with pressure or vacuum on the intake?
a. Avoids water hammer
b. Avoids losing pump's prime
c. Avoids creating backflow in line
d. Avoids damage to O-ring

____ 55. Why is some type of automatic pressure regulation for attack lines essential?
a. Prevents damage to nozzles
b. Modern nozzles intolerance of less-than-ideal pressures
c. Ensures safety of driver/operator
d. Ensures safety of nozzleperson

____ 56. NFPA 1901 requires a pressure control device to operate when the discharge pressure rises at least ____ above the set level.
a. 10 psi *(69 kPa)*
b. 20 psi *(138 kPa)*
c. 25 psi *(172.5 kPa)*
d. 30 psi *(207 kPa)*

____ 57. Which of the following statements about discharge pressure relief valves is true?
a. They rapidly reroute large quantities of water.
b. As long as pressure in a spring-loaded pilot valve's pressure chamber remains below the set level, the pilot valve stays open.
c. They react quickly to overpressure conditions but are somewhat slower to reset back to "all-closed" positions.
d. The pressure relief control valve is usually located inside the cab.

____ 58. Intake pressure relief valves generally should be set to open when intake pressure rises more than ____ above the desired operating pressure.
a. 5 psi *(35 kPa)*
b. 10 psi *(70 kPa)*
c. 15 psi *(105 kPa)*
d. 20 psi *(140 kPa)*

____ 59. Which of the following statements about pressure governors is true?
a. They may be used in connection with a throttle control, engine throttle, and/or pump discharge.
b. Engine speed is governed by water pressure on the intake side of the pump.
c. They must be attached to an auxiliary throttle.
d. Manual governors have a cavitation mode that returns the engine to idle when the intake pressure drops below 30 psi *(210 kPa)*.

____ 60. Most modern fire apparatus use ____ primers.

a. Gravity flow
b. Positive displacement
c. Exhaust
d. Vacuum

____ 61. The two purposes of a primer's oil/fluid supply are to ____.

a. Collect impurities and lubricate gears.
b. Lubricate gears and preserve metal parts.
c. Seal gaps and preserve metal parts.
d. Collect impurities and seal gaps.

____ 62. In general, gear-driven primers should operate with an engine rpm around ____.

a. 350 to 700 rpm
b. 700 to 1,000 rpm
c. 1,000 to 1,200 rpm
d. 1,200 to 1,450 rpm

____ 63. Which of the following is an advantage of exhaust primers?

a. Lightweight
b. Efficient
c. Low maintenance
d. Runs at low engine rpm

____ 64. Which of the following primer types relies on a line from the intake manifold of the apparatus engine to the intake of the fire pump?

a. Gravity flow
b. Positive displacement
c. Exhaust
d. Vacuum

____ 65. Which of the following statements about vacuum primers is ***not*** true?

a. They can draw explosive gases into the fire pump.
b. They operate best at high engine rpm.
c. They can draw water into the engine's intake manifold.
d. They are the simplest type of primer.

____ 66. Which of the following pump panel instruments is ***not*** required by NFPA 1901?

a. Weatherproof tachometer
b. Voltmeter
c. Pumping engine oil pressure indicator
d. Automatic transmission temperature gauge

____ 67. A master intake gauge is usually calibrated from ____ positive pressure.

a. 0 to 1,200 psi *(0 kPa to 8 280 kPa)*
b. 600 to 1,000 psi *(0 kPa to 6 900 kPa)*
c. 0 to 600 psi *(0 kPa to 4 137 kPa)*
d. 0 to 1,000 psi *(0 kPa to 6 900 kPa)*

____ 68. As the vacuum reading approaches ____, the pump is near its maximum capacity.

a. 20 inches *(508 mm)*
b. 15 inches *(381 mm)*
c. 12 inches *(305 mm)*
d. 10 inches *(254 mm)*

____ 69. If a pumper is equipped to supply high-pressure fog streams, the discharge pressure gauge may be calibrated up to ____.

a. 1,200 psi *(8 280 kPa)*
b. 1,500 psi *(10 350 kPa)*
c. 600 psi *(4 137 kPa)*
d. 1,000 psi *(6 900 kPa)*

____ 70. If the tachometer shows a gradual increase in the rpm required to pump the rated capacity, it indicates ____.

a. A blocked supply line
b. Wear in the pump
c. A need to lubricate the pump
d. An overheated pump

____ 71. Which of the following statements about pump pressure indicators is true?

a. They are also known as intake gauges.
b. They must be connected to the intake side of the discharge valve.
c. The master pressure gauge is the only way to be certain the pump is adjusted properly when supplying a master stream device.
d. NFPA 1901 allows flowmeter readouts to substitute for individual pressure discharge gauges.

____ 72. The most common type of water tank level indicator is a(n) ____ .

a. Series of lights
b. Sight gauge
c. Dial gauge
d. Audible warning signal

____ 73. What is an auxiliary cooler's primary function?

a. Cool pump fluid
b. Cool engine oil
c. Cool engine coolant
d. Cool pump bearings

____ 74. Which of the following statements about auxiliary cooling devices is true?

a. In an immersion-type cooler, water from the pump comes into direct contact with the engine coolant.
b. To avoid excess pressure buildup, the radiator fill valve should be fully opened.
c. The apparatus cooling system should be serviced as soon as possible after every time the radiator fill valve is used.
d. In an emergency, a water stream may be directed at the engine to maintain cooling throughout an extended incident.

Chapter 11 Operating Fire Pumps

Matching

A. Match to their definitions terms associated with operating fire pumps. Write the appropriate letters on the blanks.

____ 1. Making the fire pump operational

____ 2. Helps to cool water by allowing it to circulate through the pump and back into the tank

____ 3. In some models, gives a tally of how much water has been used

____ 4. Hose laid from the water source to the fire

____ 5. Allows a second pumper to provide additional pressure to a forward lay supply line

____ 6. Hose laid from the fire to the water source

____ 7. Pressure shown on the intake gauge when a pumper is connected to a hydrant and is not discharging water

____ 8. Noncollapsible waterway between the fire pump and static water source

____ 9. Condition where, in theory, the pump is discharging water faster than taking it in

____ 10. Prevents water flow from the sprinkler system back into the fire department connection

a. Check valve
b. Forward lay
c. Putting the pump into gear
d. Hard intake hose
e. Residual pressure
f. Cavitation
g. Four-way hydrant valve
h. Static pressure
i. Reverse lay
j. Booster cooling valve
k. Flowmeter

True/False

B. Write *True* or *False* before each of the following statements. Correct those statements that are false.

_________ 1. IFSTA recommends that apparatus wheels be chocked every time an apparatus is parked with the engine running and the driver/operator exits the cab.

____

____

_________ 2. The procedure for engaging power take-off and front mount pumps is generally the same.

____

____

_________ 3. Most driver/operators operate mainly from an external pressurized source at the vast majority of incidents.

____

____

_________ 4. The driver/operator must rely solely on previous experience and departmental SOPs when determining how much pressure to build within the pump before any lines are actually opened.

____

____

_________ 5. If the nozzles are open when the driver/operator puts the pump into operation, he must rely on experience and department SOPs to bring the bring pressure to the proper level.

____

____

_________ 6. Operating at a negative pressure from a fire hydrant increases the possibility of damage to the fire pump and water system *and* to water heaters on a municipal water supply.

____

____

_________ 7. Operating at a negative pressure when being supplied from another pumper presents the danger of the supply hose's collapsing and of causing cavitation.

____

____

_________ 8. The best hydrant is always the one closest to the fire.

____

____

_________ 9. The three-hundred-foot maximum length for a forward lay supply line working directly from hydrant pressure does not apply when using 3½-inch *(90 mm)* or larger supply hose.

____

____

_________ 10. A reverse lay is the best when boosting hydrant pressure to the supply line is necessary.

____

____

_________ 11. Reverse hose lays require the use of a four-way hydrant valve.

____

____

_________ 12. NFPA 1901 forbids discharging water from a booster line to keep the pump from overheating.

____

____

_________ 13. The first-digit method for calculating available water cannot be used with the metric system.

____

____

_________ 14. The driver/operator can rely on a water system's usual static pressure and an assumption of the reasonable flow based on hose size in order to determine how many lines can be added using the squaring-the-lines method.

____

____

_________ 15. Water hammer can break a hydrant and push it out of the ground.

____

____

_________ 16. Increasing the size of the intake hose can reduce the capacity of a pump.

____

____

_________ 17. If a drafting operation is intended to supply water directly to a fireground, the driver/operator may not have much choice when selecting a site.

____

____

_________ 18. Driver/operators should not consider small ponds or streams for drafting sources.

____

____

_________ 19. When drafting ocean water, a location may be acceptable at one time of day and unacceptable at another.

____

____

_________ 20. In an emergency, any type of water can be used for fire fighting.

____

____

_______ 21. When drafting from a static source, the driver/operator should engage the pump before leaving the cab.

_______ 22. If space around the actual drafting spot is limited, the driver/operator should position the apparatus in the actual drafting site and then have other firefighters help make the connections.

_______ 23. A two-stage pump should be in the volume position during priming.

_______ 24. Before the driver operator begins priming, all drains and valves should be open and all unused intake and discharge openings should be uncapped.

_______ 25. If the priming pump is driven by an electric motor, the exact rpm is not critical for priming.

_______ 26. If the hoselines are not ready for water after the pump is primed, another discharge or a booster line can be opened to discharge water back into the source.

_______ 27. Once the primer is shut off, even the smallest air leak can result in a loss of vacuum.

_________ 28. Operating from draft is the most demanding type of pump operation for the apparatus but is relatively easy for the operator.

_________ 29. If the inner liner of hard intake hose detaches from the hose itself and collapses, it will be immediately apparent on visual examination.

_________ 30. The water supply for a sprinkler system is designed to supply seventy percent of the sprinklers installed on the system.

_________ 31. No matter what a department's other standard operating procedures might be, upon arrival at a sprinklered property, preparations should immediately be made to supply the FDC.

_________ 32. In most cases, it is desirable to charge the sprinkler system immediately upon the fire company's arrival.

_________ 33. Generally the friction loss in the standpipe is small unless the flows are large.

Multiple Choice

C. Write the letter of the best answer on the blank before each statement.

____ 1. The parking brake control on most large, modern apparatus is a ____.

a. Hand lever
b. Push pull control
c. Foot pedal
d. Toggle switch

____ 2. The apparatus may override the parking brake system at engine speeds as low as ____.

a. 700 rpm
b. 1,000 rpm
c. 1,300 rpm
d. 1,500 rpm

____ 3. What should the driver/operator do if the indicator light does not illuminate after following the procedure to engage a PTO or front-mount pump into gear?

a. Shut down immediately.
b. Visually check pump operation.
c. Repeat the procedure.
d. Manually engage the pump.

____ 4. For stationary operation of a PTO or front-mounted pump, the speedometer should show a speed ____.

a. Of 0 mph *(0 km/h)*
b. Slightly above 0 mph *(0 km/h)*
c. Of more than 5 mph *(8 km/h)*
d. Between 5 and 10 mph *(8 km/h and 16 km/h)*

____ 5. Which of the following statements about engaging midship transfer driven pumps is true?

a. They require that the truck transmission be in neutral.
b. The drive transmission should be in first gear.
c. In older apparatus the speedometer will remain at 0 mph *(0 km/h)* after the pump is engaged.
d. If the speedometer shows 0 mph *(0 km/h)* after the pump is engaged, the driver/operator should rely on the visual pump shift indicator.

____ 6. What should the driver/operator do if the pump shift indicator light remains illuminated after disengaging a midship transfer driven pump.

a. Repeat the procedure.
b. Carefully attempt to put the apparatus in motion.
c. Visually inspect pump operation.
d. Shut down immediately.

____ 7. Which of the following statements about engaging auxiliary-engine driven fire pumps is true?

a. They must be placed into gear from neutral after the engine is running.
b. If the auxiliary engine is running, the water supply to the pump must be opened.
c. NFPA 1901 prohibits using pull cord starters for auxiliary engines.
d. Electric starter switches must be mounted directly on the auxiliary engine.

____ 8. Which of the following statements about putting the pump into operation is true?

a. The tank-to pump valve should locked in the open position before the wheels are chocked.

b. In most cases a multistage pump should be in a parallel position.

c. If the pump will be required to furnish more than 50 percent of its rated capacity, it should initially be set to the volume position.

d. After opening the tank-to-pump valve the driver should return to the cab and increase engine rpm.

____ 9. What should the driver/operator do if none of the attack lines are ready to be charged by the time the pump pressure has built up?

a. Partially open the pump-to-tank line.

b. Throttle down to maintain minimum pressure.

c. Open the discharge valve just enough to maintain proper pressure.

d. Partially open an unused discharge valve.

____ 10. Which of the following statements about making the transition from the onboard tank to an external water supply is true?

a. The supply line should be connected to an intake with an open gate valve.

b. The pump-to-tank valve should be used to refill the water tank only when hoselines are shut down.

c. The tank-to-pump valve should be closed before the intake valve is opened.

d. The bleeder valve on the gated intake line should be opened to purge air from the supply line.

____ 11. If the discharge pressure is increased too much, the intake pressure from the supply source may be reduced ____.

a. To a minimum equal to discharge pressure

b. To a minimum equal to the difference between intake pressure

c. To a minimum of 0 psi *(0 kPa)*

d. Below 0 psi *(0 kPa)*

____ 12. As a standard practice it is not desirable to reduce the incoming pressure from a hydrant or supply pumper below ____.

a. 10 psi *(70 kPa)*

b. 15 psi *(105 kPa)*

c. 20 psi *(140 kPa)*

d. 25 psi *(175 kPa)*

____ 13. Which of the following is the ***best*** hydrant?

a. Large water main in a grid

b. Dead end hydrant

c. Medium water main in a grid

d. Hydrant on a large single line

____ 14. Which of the following is the ***worst*** hydrant?

a. Large water main in a grid

b. Dead end hydrant

c. Medium water main in a grid

d. Hydrant on a large single line

____ 15. Which of the following NFPA standards covers flow testing and marking of hydrants?

a. 291

b. 921

c. 1901

d. 9000

____ 16. A class B hydrant is ____.

a. Green and flows 500 – 999 gpm *(1 900 L/min to 3 780 L/min)*

b. Green and flows 1, 500 gpm *(5 680 L/min)*

c. Orange and flows 1,000 – 1,499 gpm *(3 785 L/min to 5 675 L/min)*

d. Orange and flows 500 – 999 gpm *(1 900 L/min to 3 780 L/min)*

____ 17. Which of the following statement about four-way hydrant valves is true?

a. The first pumper connects to the large-diameter pumper intake connection.

b. The supply line to the first pumper should not be opened until the discharge line from the second pumper to the hydrant is primed.

c. When pressure from the second pumper overcomes pressure from the hydrant, a clapper valve in the four-way hydrant valve closes.

d. When the four-way hydrant is fully operational, the supply line to the first pumper is being fed from both the hydrant and the second pumper.

____ 18. When a 2½- or 3-inch *(65-mm or 77-mm)* hoseline is supplying a pumper directly off hydrant pressure, it should be no longer than ____.

a. 150 feet *(45 m)*

b. 200 feet *(60 m)*

c. 250 feet *(75 m)*

d. 300 feet *(90 m)*

____ 19. Which of the following is an advantage of a reverse lay?

a. It facilitates scene size up before making the hydrant connection.

b. It speeds starting the initial attack.

c. It allows the pump operator to perform other essential fire location activities.

d. It can be used regardless of how hose with threaded couplings is laid in the apparatus hose bed.

____ 20. Which of the following statements about getting water into the pump is true?

a. The tank to pump valve must be open if the intake is not equipped with a shutoff valve.

b. If the intake is equipped with a shutoff valve, the driver/operator may charge the intake line and bleed off air while pumping water from the tank.

c. A dry-barrel hydrant should be opened until it easily stops turning and then tightened in the open position another quarter-turn.

d. As the pump fills with water from a static water source, the master intake gauge indicates static pressure in the water supply.

____ 21. Which of the following statements about putting the pump in service is true?

a. The driver/operator should not engage the pump drive system before leaving the cab if more than a few minutes will pass before water is introduced to the pump.

b. The pump on a hydrant or in the middle of a relay usually should be set to the series position.

c. If the master intake gauge drops below 20 psi *(40 kPa)* the driver operator should increase the throttle.

d. The water tank should not be allowed to fill and overflow through the tank vent.

____ 22. What indicates how much more water a hydrant can supply when it is discharging water?

a. Residual pressure

b. Static pressure

c. The average of residual and static pressure

d. The difference between residual and static pressure

____ 23. When using the percentage method to determine additional water available from a hydrant, how many lines with the same flow as the first can be added if the drop is 11 – 15 percent?

a. Three

b. Two

c. One

d. None

____ 24. If the difference in static and residual pressure is greater than 25 percent, how much additional water is available?

a. Three times the amount being delivered

b. Two times the amount being delivered

c. The amount being delivered

d. Less than the amount being delivered

____ 25. When using the first-digit method for calculating available water, a psi drop equal to or less than the first digit of the static pressure multiplied by ____ indicates that one additional line of equal flow may be added.

a. One
b. Two
c. Three
d. Four

____ 26. The squaring-the-lines method of calculating available water involves multiplying the square of the number of lines currently in operation by ____.

a. Static pressure
b. Residual pressure
c. The original pressure drop
d. The pressure drop after adding the preceding line

____ 27. When using the squaring the lines method of calculating available water, the pressure drop in the system is ____ each time another line the same size as the first line is added.

a. Halved
b. Doubled
c. Tripled
d. Quadrupled

____ 28. Which of the following statements about shutting down the hydrant operation is true?

a. The driver/operator should take the pressure control device out of service and then gradually slow engine rpm.
b. The last turn of the operating nut opens the drain holes in the base of a dry barrel hydrant.
c. The driver/operator should slowly and smoothly close the discharge valves after placing the drive transmission into neutral.
d. The operator can visually verify that a dry barrel hydrant is draining after the hydrant is closed and the caps are in place.

____ 29. What principle allows a pumper to draw water from a static source below the pump?

a. Adherence of drops of water
b. Gravitational pull
c. Pressure difference between the atmosphere and the pump
d. Newton's first law of motion

____ 30. Which of the following is ***not*** a significant factor in pressure loss?

a. Volume of water
b. Hose appliance
c. Inertia of water
d. Atmospheric pressure at intake

____ 31. Atmospheric pressure at sea level is about ____.

a. 14.7 psi *(100 kPa)*
b. 12.7 psi *(86 kPa)*
c. 10.0 psi *(68 kPa)*
d. 0 psi *(0 kPa)*

____ 32. Atmospheric pressure decreases ____ for each 1,000 feet *(305 m)* of elevation gain.

a. 0.50 psi *(3.5 kPa)*
b. 0.75 psi *(5.25 kPa)*
c. 1.00 psi *(7.0 kPa)*
d. 1.25 psi *(8.75 kPa)*

____ 33. Most pumps deliver a maximum of approximately ____ of mercury.

a. 11 inches *(280 mm/hg)*
b. 17 inches *(435 mm/hg)*
c. 19 inches *(485 mm/hg)*
d. 22 inches *(560 mm/hg)*

____ 34. The size and number of suction hoses for a pump with a rated capacity of 1,000 gpm *(4 000 L/min)* is ____.

a. One 4½-inch *(115-mm)* hose
b. One 5-inch *(125-mm)* hose
c. Two 5-inch *(125-mm)* hoses
d. One 6-inch *(150-mm)* hose

____ 35. What is the ***best*** indication of cavitation?

a. Fluctuating hose streams
b. Lack of pressure gauge reaction to throttle changes
c. Popping or sputtering at the nozzle
d. Sound like gravel passing through the pump

____ 36. Which of the following statements about cavitation is true?

a. Cavitation rarely results from inadequate piping from the water tank to the pump.
b. Cavitation damage can be especially severe when the pump is supplying short attack lines at high pressure.
c. Cavitation most often occurs during drafting operations.
d. When cavitation occurs, increasing pump rpm will increase discharge pressure.

____ 37. What is the first consideration in establishing a successful drafting operation?

a. Size of suction hose
b. Location of fire
c. Positioning the apparatus
d. Selecting the drafting site

____ 38. Which of the following is the most important factor in selecting a drafting site?

a. Amount of water available
b. Type of apparatus
c. Cost of water
d. Accessibility of water

____ 39. What minimum depth of water over a traditional strainer is generally needed for a pumper to approach its rated capacity?

a. 12 inches *(305 mm)*
b. 18 inches *(455 mm)*
c. 24 inches *(610 mm)*
d. 30 inches *(765 mm)*

____ 40. Which of the following is the best solution when drafting from a shallow stream?

a. Build a dam.
b. Dig out the bottom.
c. Build a dam ***and*** dig out the bottom.
d. Use a floating strainer.

____ 41. A disadvantage of the floating strainer is that ____.

a. It may create a whirlpool.
b. It must not be allowed to float free in the water.
c. Its restricted intake may limit the pumper's ability to reach its rated capacity.
d. The holes on its top are easily obstructed.

____ 42. Low-level strainers are a good choice when drafting from a ____ .

a. Shallow pond
b. Fast moving stream
c. Swimming pool
d. Spring

____ 43. Which of the following statements about water temperature is true?

a. It is not a consideration when choosing a type of water to draft.
b. Water may be 5 – 10°F *(3°C to 5°C)* warmer at a depth of several feet than at the surface.
c. Heated water is preferable to water that is close to freezing.
d. Heated water is most commonly found at discharges from factories.

____ 44. What is the most common type of contamination when drafting nonpotable water?

a. Salt
b. Sulfur
c. Algae
d. Dirt or sand

____ 45. What happens as the amount of lift increases?

a. Elevation pressure increases.
b. Friction loss is easier to overcome.
c. Pump capacity increases.
d. Contaminants become more likely.

____ 46. Pumps meeting NFPA and Underwriter's Laboratories requirements are rated to pump their capacity at ____ of lift.

a. 0 feet *(0 m)*
b. 10 feet *(3 m)*
c. 20 feet *(6 m)*
d. 30 feet *(9 m)*

____ 47. The maximum lift considered reasonable for most fire department pumpers is ____.

a. 0 feet *(0 m)*
b. 10 feet *(3 m)*
c. 20 feet *(6 m)*
d. 30 feet *(9 m)*

____ 48. The key to making connections airtight among the strainer and needed sections of hard intake hose is ____.

a. Using an approved coupling lubricant
b. Using a rubber mallet to tighten all connections
c. Connecting the strainer last
d. Getting each section of hose in line with the other before turning the coupling

____ 49. What should firefighters do to ease positioning the strainer and handling the hard intake hose after it is connected?

a. Use strainers with built-in wheels.
b. Fasten a rope to the end of the strainer.
c. Remove gaskets before making connections.
d. Use the apparatus to drag the hard intake hose and strainer.

____ 50. Which of the following statements about connecting to the pump is true?

a. It is usually easier to make all connections before putting the strainer in the water.
b. If a butterfly valve is normally connected to the pump intake, it should be removed before connecting the intake hose to the apparatus.
c. Placing a shovel under a barrel strainer is not , under any circumstances, an acceptable method of protecting the strainer.
d. Hard intake hose can be routed over obstructions that are above the level of the pump.

____ 51. Most priming pumps are intended to work best when the engine speed is between ____.

a. 300 and 500 rpm
b. 700 and 950 rpm
c. 1,000 and 1,200 rpm
d. 1,250 and 1,500 rpm

____ 52. As air is evacuated from the pump, the master intake gauge should register a vacuum reading of ____ for each 1 foot *(0.3 m)* of lift.

a. 1 inch *(25 mm)*
b. 2 inches *(50 mm)*
c. 6 inches *(150 mm)*
d. 1 foot *(300 mm)*

____ 53. If the primer discharges a steady stream of water under the apparatus during priming, it indicates ____.

a. Normal operation
b. Inadequate lift
c. Leaking gaskets
d. Dangerously high intake pressure

____ 54. How long does the entire priming action usually take?

a. 45 to 60 seconds
b. 20 to 40 seconds
c. 15 to 30 seconds
d. 10 to 15 seconds

____ 55. What is the most common cause of inability to prime?

a. Obstructed hard intake hose
b. Clogged strainer
c. Closed intake gate valve
d. Air leak

____ 56. What is the most common cause of an air leak during priming?

a. Broken hard intake hoses
b. Damaged gaskets
c. Open drains and valves
d. Insufficient fluid in the priming reservoir

____ 57. After the pump is primed, the throttle setting should be increased to boost discharge pressure to somewhere between ____.

a. 75 and 125 psi *(525 kPa and 875 kPa)*
b. 50 and 100 psi *(350 kPa and 700 kPa)*
c. 100 and 150 psi *(700 kPa and 1 050 kPa)*
d. 125 and 175 psi *(875 kPa and 1 225 kPa)*

____ 58. When operating from draft, any significant deviation from normal engine temperature is a signal that ____.

a. Engine speed is too high.
b. The pump was not properly primed.
c. Another pumper should be used for drafting.
d. Water is not routed correctly.

____ 59. What is the most common problem when operating from draft?

a. Air leak on the intake side of the pump
b. Whirlpool allowing air to enter the pump
c. Air leak due to defective packing in the pump
d. Air leak on the discharge side of the pump

____ 60. Which of the following statements about operating the pump from draft is true?

a. An immediate cure to cavitation is to increase engine rpm until the pressure drops.

b. A gradual increase in the vacuum reading with no change in the flow rate indicates a blockage is developing.

c. Defective packing can be temporarily fixed with a heavy coat of grease while the pump is operating.

d. The most common place for a blockage to develop is at the intake valve.

____ 61. After the pump is drained and the connections have been removed, the driver/operator should operate the mechanical primer ____.

a. For one minute

b. Until the pump lubrication indicator light goes off

c. Until the priming pump has refilled with water from the pump-to-tank connection

d. Until primer oil or fluid comes out of the discharge for the priming pump

____ 62. Fire department connections consist of a siamese with at least two 2½-inch *(65 mm)* female connections or ____.

a. One 4-inch *(100 mm)* female connection

b. One large sexless connection

c. Three 2-inch *(50 mm)* female connections

d. One 2-inch *(50 mm)* female connection and one 3-inch *(77 mm)* sexless connection

____ 63. As a general rule of thumb, one 1,000 gpm *(4 000 L/min)* rated pumper should supply the FDC for every ____ sprinklers estimated to be flowing.

a. 25

b. 50

c. 75

d. 100

____ 64. If the driver/operator does not know the recommended pressure for charging a particular sprinkler system, a general rule of thumb is to discharge ____ into the sprinkler system.

a. 100 psi *(700 kPa)*

b. 150 psi *(1 050 kPa)*

c. 200 psi *(1 400 kPa)*

d. 250 psi *(1 750 kPa)*

____ 65. Which of the following statements about sprinkler system control valves is true?

a. When possible, the first arriving unit should ask the property owner or responsible party to ensure that they are open.

b. Where it is known that construction in a building has affected sprinklers, the control valve should immediately be opened.

c. The firefighter or crew assigned to check control valves should carry a flashlight and portable radio.

d. Control valves and the pump should be located in different rooms.

____ 66. Which of the following statements about standpipe systems is true?

a. 2½- *(65 mm)* or 3-inch *(77 mm)* connections are provided on each floor.

b. Fire personnel should use house or standpipe lines when available.

c. Standpipe and sprinkler systems work off of the same fire department connection.

d. Wet standpipes have an FDC that should be supplied with water under pressure to supplement the system's water supply.

____ 67. The driver/operator should add approximately ____ to the desired nozzle pressure for each floor above the standpipe connection that will have operating fire streams.

a. 5 psi *(35 kPa)*

b. 10 psi *(70 kPa)*

c. 15 psi *(105 kPa)*

d. 20 psi *(140 kPa)*

____ 68. Pump discharge pressures above ____ are generally not encouraged.

a. 100 psi *(700 kPa)*

b. 150 psi *(1 050 kPa)*

c. 200 psi *(1 400 kPa)*

d. 250 psi *(1 750 kPa)*

____ 69. The elevation pressure for a standpipe system equipped with pressure-reducing valves must be based on the ____.

a. Square of the standpipe's diameter

b. Total height of the standpipe or zone being used

c. Actual height of the attack nozzle

d. Height of the standpipe multiplied by the square of its diameter

____ 70. If a fire department connection is totally unusable, the standpipe riser ____.

a. Can be charged at the first floor level by attaching a double female to a hose valve at the first floor level.

b. Cannot be used.

c. Can be charged by attaching a double male to a hose valve one floor below the fire attack.

d. Can be supplied by opening a check valve in the sprinkler system, if available.

Chapter 12 Static Water Supply Sources

Matching

A. Match to their definitions terms associated with static water supply sources. Write the appropriate letters on the blanks.

____ 1. Process of raising water from a static source to supply a pumper

____ 2. Elevation difference between static water source and center of pump

____ 3. Difference between atmospheric pressure and pressure in hose and pump

____ 4. Greatest height to which any amount of water may be raised through a hard intake hose to the pump

____ 5. Height a column of water may be lifted in sufficient quantity to provide a reliable fire flow

____ 6. Sum of the pump discharge pressure and intake pressure correction

____ 7. Underground water storage receptacle

____ 8. Manmade impoundment with the same characteristics of a pond or small lake

a. Net discharge pressure

b. Maximum lift

c. Lift

d. Cistern

e. Peak useable force

f. Drafting

g. Ground reservoir

h. Partial vacuum

i. Dependable lift

True/False

B. Write *True* or *False* before each of the following statements. Correct those statements that are false.

________ 1. Driver/operators in jurisdictions with pressurized water supplies do not need to be familiar with static water supply sources.

____

____

______ 2. Fire service equipment cannot create total vacuums.

______

______

______ 3. The angle of the intake hose affects the height of possible lift.

______

______

______ 4. As the pump approaches maximum lift, the tolerable amount of friction loss in the intake hose is so low that the volume of water available for the pump is too low to be of practical value.

______

______

______ 5. All fire department pumping apparatus have their fire pumps rated when drafting from a lift of 10 feet *(3 m)* through 20 feet *(6 m)* of hard intake hose.

______

______

______ 6. The two primary factors when evaluating a natural static water supply source are adequacy and quality of water.

______

______

______ 7. Fire departments should not use public boat launching facilities for drafting sites.

______

______

______ 8. When an inaccessible water supply is the closest source, portable pumps should be positioned at that source.

______

______

______ 9. If the apparatus is in an effective position and pumping water when it begins sinking in soft ground, operations should continue until the need for water has ended.

______

______

_______ 10. Some cisterns are placed specifically for fire department use.

____

____

_______ 11. Technically, elevated tanks are not a static source of water.

____

____

_______ 12. If an apparatus must be driven onto private property to access a swimming pool, floating or stationary portable pumps should be used to pump the water to the apparatus.

____

____

_______ 13. Most swimming pools contain enough water to extinguish a fire in a single-family dwelling.

____

____

_______ 14. Some larger indoor swimming pools may be equipped with a dry-hydrant type connection.

____

____

_______ 15. Pumps do not have to be flushed after pumping from a swimming pool.

____

____

Multiple Choice

C. Write the letter of the best answer on the blank before each statement.

____ 1. In the US system of measurement, a total vacuum at sea level would allow a pumper to lift water ____ feet.

a. 28.7
b. 31.46
c. 33.8
d. 38.8

____ 2. In the US system atmospheric pressure decreases by about ____ psi for every 1,000 feet of altitude.

a. 0.25
b. 0.50
c. 0.75
d. 1.00

____ 3. In the metric system of measurement, a total vacuum at sea level would allow a pumper to lift water ____ m.

a. 5
b. 10
c. 15
d. 20

____ 4. In the metric system atmospheric pressure decreases by about ____ kPa for every 100 m of altitude.

a. 1
b. 2
c. 5
d. 10

____ 5. In most circumstances, maximum lift is no more than ____.

a. 10 feet *(3 m)*
b. 15 feet *(4.5 m)*
c. 20 feet *(6 m)*
d. 25 feet *(7.5 m)*

____ 6. What is the maximum height to which a pump can lift water when the intake gauge reads 11 inches of mercury?

a. 11.00 feet
b. 1.13 feet
c. 12.43 feet
d. 12.77 feet

____ 7. What is the maximum height to which a pump can lift water when the intake gauge reads 190 mm of mercury?

a. 1.90 m
b. 2.65 m
c. 19.0 m
d. 24.8 m

____ 8. According to Table 12.1a, a pump with a rated capacity of 1,000 gpm will discharge at least ____ gpm when being supplied by a 6-inch hose at 14 feet of lift if pump pressure is 150 psi.

a. 1,105
b. 998
c. 870
d. 1,045

____ 9. According to Table 12.1b, a pump with a rated capacity of 4 000 L/min will discharge at least ____ L/min when being supplied by a 125-mm hose at 5.5 m of lift if pump pressure is 1 000 kPa.

a. 2 479
b. 2 120
c. 2 536
d. 1 874

____ 10. Every fire pump in good repair should have a dependable lift of at least ____.

a. 10 feet *(3 m)*
b. 14.7 feet *(4.5 m)*
c. 35 feet *(10 m)*
d. 17.1 feet *(5.25 m)*

____ 11. At a 15 foot *(4.5 m)* lift, a pump can deliver only ____ percent of its rated capacity.

a. 50
b. 60
c. 70
d. 80

____ 12. Using Table 12.2a, determine net pump discharge pressure for a 1,500 gpm pumper if the required lift is 14 feet through 30 feet of 6-inch hose and pump discharge pressure is 135 psi.

a. 10.87 psi
b. 124.13 psi
c. 145.87 psi
d. 163.63 psi

____ 13. Using Table 12.2b, determine net pump discharge pressure for a 3 000 L/min pumper if the required lift is 3.5 m through 6 m of 115-m intake hose and pump discharge pressure is 900 kPa.

a. 60.5 kPa
b. 839.5 kPa
c. 939.5 kPa
d. 960.5 kPa

____ 14. A 6-foot wide and 3-foot deep stream that is flowing approximately 25 feet per minute will provide about ____ gpm.

a. 3,375
b. 450
c. 2,996
d. 3,825

____ 15. A 2.5-m wide and 0.75-m deep stream that is flowing approximately 2 meters per minute will provide about ____ L/min.

a. 1 875
b. 375
c. 3 750
d. 5 625

____ 16. The rule of thumb for evaluating pond and small lake capacity is that every foot *(0.3 m)* of depth for an area of 1 acre *(0.4 ha)* provides ____ for 5 hours.

a. 750 gpm *(3 000 L/min)*
b. 1,000 gpm *(4 000 L/min)*
c. 1,250 gpm *(5 000 L/min)*
d. 1,500 gpm *(6 000 L/min)*

____ 17. Which of the following statements about drafting depth is true?

a. A good rule of thumb is that a barrel-type strainer requires a minimum depth of 3 feet *(1 m)*.
b. Floating strainers allow safe drafting from water down to a 1- to 2-inch *(25 mm to 50 mm)* depth.
c. Low-level strainers are most commonly used to draft from creeks and streams.
d. A ladder and salvage cover can be used to raise the depth of a small, fast running stream.

____ 18. Which of the following statements about silt and debris is true?

a. The strainer should rest just on the bottom of the water supply source.
b. Only roof ladders should be used to support a strainer.
c. Dry hydrants avoid problems with wet ground approaches but creates slow set-up time.
d. The pump must be flushed with clean fresh water after every operation using salt water or dirty water.

____ 19. Which of the following statements about operations in freezing weather is ***not*** true?

a. To overcome surface freezing, barrels filled with antifreeze solution may be floated on the water before it freezes.

b. Firefighters should not begin work on ice until provisions for making an ice rescue are en route.

c. Firefighters should not use wooden plugs or plastic garbage cans to drive through the ice to draft water.

d. Firefighters working on ice may not need to wear personal flotation devices if they are certain the water below them is less than chest deep.

____ 20. Cistern capacities commonly range up to ____.

a. 25,000 gallons *(100 000 L)*

b. 50,000 gallons *(200 000 L)*

c. 75,000 gallons *(300 000 L)*

d. 100, 000 gallons *(400 000 L)*

____ 21. Which of the following statements about ground reservoirs is true?

a. They typically hold many millions of gallons *(liters)* of water.

b. They are usually less accessible than regular ponds or lakes.

c. The fire department should not use dry hydrants at these sites.

d. Strainers are not required at these sites.

____ 22. The capacity of a swimming pool that is 25 feet wide and 35 feet long with an average depth of 4.5 feet is approximately ____ gallons.

a. 29,500

b. 6,600

c. 3,900

d. 1,000

____ 23. The capacity of a round swimming pool that is 15 m in diameter and 1.5 m deep is approximately ____ liters.

a. 1 060 000

b. 85 000

c. 265 000

d. 175 000

____ 24. Which of the following statements about agricultural irrigation systems is true?

a. Fire departments should not use them as water sources.

b. Some may flow up to 1,000 gallons *(4 000 L)* per minute.

c. Canal systems are more common than piped systems.

d. All state and local codes now require agricultural irrigation systems to have fittings that fire departments can use without special tools or adapters.

Chapter 13 Relay Pumping Operations

Matching

A. Match to their definitions terms associated with relay pumping operations. Write the appropriate letters on the blanks.

____ 1. Uses a pumper at the water supply source to pump water under pressure through one or more hoselines to the next pumper in line

____ 2. Is connected to water supply at beginning of relay operation

____ 3. Receives water from another pumper and then supplies water to the next pumper or an attack pumper

____ 4. Pumper located at the fire scene

____ 5. Allows air to escape from the system as hoses are charged

____ 6. Three to five pumpers with large capacity fire pumps and, usually, large diameter hose

a. Relay relief valve
b. Bleeder valve
c. Source pumper
d. Relay task force
e. Attack pumper
f. Relay pumping operation
g. Relay pumper

True/False

B. Write *True* or *False* before each of the following statements. Correct those statements that are false.

________ 1. Hose tenders usually carry a mile or more of large diameter (4-inch [100 mm] or larger) hose.

____

____

________ 2. Relay pumping operations require large diameter hose.

____

____

_________ 3. Bleeder valves may be located directly in the intake piping to the pump itself.

____

____

_________ 4. If low flow rates allow greater spacing between pumpers than the amount of hose they carry, some pumpers may lay their hose without actually participating in the pumping process.

____

____

_________ 5. The two basic designs for relay pumping operations are minimum distance and maximum distance.

____

____

_________ 6. Different size hoses can be used in a relay as long as the distances between pumpers are varied accordingly.

____

____

_________ 7. The attack pumper does not need a dump line.

____

____

_________ 8. Manual pressure control is preferable to relying on automatic pressure control devices in a relay operation.

____

____

_________ 9. Too much radio usage in establishing a relay can hamper not only fire fighting activities but also efforts to establish the water supply.

____

____

_____ 10. Radio equipped ambulances or utility units that are not otherwise occupied can be used to establish communications throughout a relay when all of the pumpers do not have a common radio channel to work on.

____

____

_____ 11. Relay operations should be shut down in order from the supply source to the fire scene.

____

____

Multiple Choice

C. Write the letter of the best answer on the blank before each statement.

____ 1. Which of the following apparatus types is most often used in fire department relay operations?

a. Specially equipped pumper
b. Pumper-tanker
c. Aerial apparatus
d. Standard pumper

____ 2. Intake pressure relief valves should be set within ____ of the static pressure of the water system or ____ of the discharge pressure of the previous pumper in the relay.

a. 10 psi *(70 kPa)*; 10 psi *(70 kPa)*
b. 10 psi *(70 kPa)*; 20 psi *(140 kPa)*
c. 5 psi *(35 kPa)*; 10 psi *(70 kPa)*
d. 15 psi *(105 kPa)*; 15 psi *(105 kPa)*

____ 3. Which of the following appliances allows later-arriving pumpers to hook up without interrupting operations after a relay is operating?

a. Relay valve
b. Water shutter
c. Wye
d. Discharge manifold

____ 4. Which of the following appliances breaks down the LDH into two or more hoselines that may then be connected to the attack pumpers?

a. Relay valve
b. Water shutter
c. Wye
d. Discharge manifold

___ 5. Which of the following statements about relay pumping operational considerations is true?

a. Increasing the size of hoses is generally the most practical way of increasing the flow of an operational hoseline.

b. The largest hose available should be laid at the beginning of the operation.

c. All fire department pumpers are rated to pump their maximum volume capacity at a net pump discharge pressure of 200 psi *(1 333 kPa)*.

d. Operating the pump at a pressure higher than 150 psi *(1 000 kPa)* increases the pumper's volume capability proportionately.

___ 6. The maximum operating pressure of a 3½-inch *(90-mm)* relay hose manufactured after 1987 is ____.

a. 150 psi *(1 030 kPa)*
b. 185 psi *(1 275 kPa)*
c. 200 psi *(1 380 kPa)*
d. 250 psi *(1 720 kPa)*

___ 7. The annual service test pressure of a 6-inch *(150-mm)* relay hose manufactured before 1987 is ____.

a. 150 psi *(1 030 kPa)*
b. 185 psi *(1 275 kPa)*
c. 200 psi *(1 380 kPa)*
d. 250 psi *(1 720 kPa)*

___ 8. If a department orders hose designed for pressures above the NFPA minimums, the hose should be pumped at pressures no greater than ____ percent of the annual service test pressure.

a. 75
b. 80
c. 85
d. 90

___ 9. The maximum distance for a hose relay section flowing 750 gpm through two 2½-inch lines is ____ feet.

a. 640
b. 160
c. 1,050
d. 1,440

___ 10. The maximum distance for a hose relay flowing 4 000 L/min through one 125 mm hose is ____ meters.

a. 270
b. 1 050
c. 110
d. 590

___ 11. At 200 psi *(1 350 kPa)* pumpers are rated to flow ____ percent of their maximum volume.

a. 50
b. 70
c. 80
d. 95

___ 12. Using the maximum distance relay method, how many pumpers are needed to flow 500 gpm through one 4-inch hose in a relay operation that is 5,000 feet long?

a. Two
b. Three
c. Four
d. Five

____ 13. How many pumpers are needed to flow 3 000 L/min through two 65-mm hoses in a relay operation that is 700 meters long.

a. Two
b. Three
c. Four
d. Five

____ 14. Which of the following statements about a constant pressure relay is true?

a. It requires complicated calculations on the emergency scene.
b. The driver/operator maintains the desired pressure by opening and closing additional discharge valves.
c. Ordinarily, the relay should maintain a constant pressure of 175 psi *(1 200 kPa)*.
d. When a decrease in flow is required, the attack pumper throttles down.

____ 15. A reasonable estimate of flow in a 750-foot *(225-m)* relay section using one 2½-inch *(65-mm)* hose and one 3-inch *(77-mm)* hose is ____.

a. 643 gpm *(2 434 L/min)*
b. 830 gpm *(3 142 L/min)*
c. 1,017 gpm *(3 849 L/min)*
d. 1,312 gpm *(4 966 L/min)*

____ 16. Which of the following statements about putting a relay into operation is true?

a. As a rule, the first arriving pumper should be placed at the source.
b. The dump line discharge valve should be left open until after the desired discharge pressure is reached.
c. NFPA 1901 forbids using the apparatus water tank to fill the hoselines.
d. The relay pumper should not place the pump in gear if it will have to wait only a few minutes for water.

____ 17. Intake pressure in a relay operation should be ____.

a. 10 to 20 psi *(75 kPa to 150 kPa)*
b. 20 to 30 psi *(150 kPa to 225 kPa)*
c. 30 to 40 psi *(225 kPa to 300 kPa)*
d. 40 to 50 psi *(300 kPa to 375 kPa)*

____ 18. If a pumper is equipped with a pressure relief valve, it should be set at ____ above the static pressure of the water system or the discharge pressure of the previous pumper in the relay.

a. 10 psi *(70 kPa)*
b. 15 psi *(105 kPa)*
c. 20 psi *(140 kPa)*
d. 25 psi *(175 kPa)*

____ 19. If an attack pumper is equipped with a readily adjustable intake relief valve, it should be set between ____.

a. 0 and 25 psi *(0 kPa and 225 kPa)*
b. 25 and 50 psi *(225 kPa and 350 kPa)*
c. 50 and 75 psi *(350 kPa and 525 kPa)*
d. 75 and 100 psi *(525 kPa and 700 kPa)*

____ 20. No adjustments should be necessary when operating a relay as long as intake pressure does not increase above ____.

a. 25 psi *(225 kPa)*
b. 50 psi *(350 kPa)*
c. 75 psi *(525 kPa)*
d. 100 psi *(700 kPa)*

Identify

D. Identify the following abbreviations associated with relay pumping operations. Write the correct interpretation before each.

________________________________ 1. MDH

________________________________ 2. LDH

Chapter 14 Water Shuttle Operations

Matching

A. Match to their definitions terms associated with water shuttle operations. Write the appropriate letters on the blanks.

____ 1. Process in which tenders deliver their load of water to the emergency scene, travel to a filling site, reload with water, and then return to the emergency scene to dump again

____ 2. Location of pumper at a water supply source

____ 3. Transfers water from source to tenders

____ 4. Location of pumper at portable water tanks

____ 5. Drafts water from portable tanks to supply attack lines or attack pumper

____ 6. Apparatus that carries at least 1,000 gallons *(3 785 L)* of water

____ 7. Tender with 750 gpm *(3 000 L/min)* or larger fire pump

____ 8. Large tank discharge

____ 9. Uses a small diameter in-line discharge inserted into the piping of the large tank discharge

____ 10. Firefighter assigned to handle a tender fill line

____ 11. Tender pumping water from its tank into an attack pumper's fire pump

____ 12. Very large tender parking immediately next to an attack pumper

a. Dump valve

b. Dump site

c. Direct pumping operation

d. Pumper tanker

e. Nurse tender operation

f. Water shuttle

g. Gravity dump

h. Fill site pumper

i. Tender

j. Fill site

k. Make and break personnel

l. Dump site pumper

m. Jet-assisted dump

True/False

B. Write *True* or *False* before each of the following statements. Correct those statements that are false.

_________ 1. The dump site pumper should not be used for "double duty" as an attack pumper.

____

____

_________ 2. NFPA 1901 requires that tenders be designed to be filled at a rate of at least 1,250 gpm *(4 730 L/min).*

____

____

_________ 3. All-wheel-drive pumpers often have front-mounted pumps that allow them to be nosed into static water supply sources.

____

____

_________ 4. Tenders that are used solely for water shuttle operations do not require a fire pump if they have a suitable gravity dump system.

____

____

_________ 5. Milk trucks should not be converted for use as tenders.

____

____

_________ 6. Travel time between dump and fill sites is a fairly fixed constant.

____

____

_________ 7. Longer handling times may negate any advantages of large diameter hose's increased flow rates for fill site operations.

____

____

_________ 8. Jet-assisted dumps are less expensive and easier to operate than large gravity dumps.

____

____

_________ 9. Inadequate venting during dumping operations can collapse the tank.

____

____

_________ 10. The dump site, fill site, and route of travel for tenders is best planned by the first responding apparatus en route to the scene.

____

____

_________ 11. The best location for a dump site is always the front and center of the incident.

____

____

_________ 12. The closest *suitable* water supply source may not be the actual closest water supply source.

____

____

_________ 13. A shuttle route that allows apparatus to drive back and forth on the same street from the fill site to the dump site is usually the most desirable.

____

____

_________ 14. Any experienced member of the fire department can be put in charge of a fill site or dump site.

____

____

_________ 15. IFSTA recommends using a pumper to fill tenders at all fill site operations, regardless of whether a hydrant or static source is used.

____

____

_________ 16. Regardless of the source, some type of small discharge line within view of the driver/operator should be continuously flowed from the fill site pumper.

____

____

_________ 17. One advantage of filling tenders by using overhead pipes and a portable manifold is that positioning the tenders is fast and easy.

____

____

_________ 18. Filling a tanker through the top with a portable fill device or an open hose butt is not recommended.

____

____

_________ 19. Filling two tenders at a time is generally the recommended water shuttle procedure.

____

____

_________ 20. Tenders should move into and out of position only when they are signaled to do so.

____

____

_________ 21. A tanker should pump its water into a portable tank only if it is not equipped with a dump valve.

____

____

_________ 22. Baffles must allow water movement at the bottom of the tank and be airtight at the top.

____

____

_________ 23. Pumping and dumping a tanker's load at the same time lowers dumping time significantly.

____

____

_________ 24. The dump site pumper should prime its pump and start water flowing as soon as the water in the portable tank is deep enough for drafting.

____

____

_________ 25. In multiple portable tank operations, water may be transferred from each tank to the next.

____

____

_________ 26. If four tanks are to be connected using jet siphons, the dump site pumper should be able to supply the hoses to the jet siphons.

____

____

_________ 27. Before shutting down a dump site, it is generally a good idea to make sure that the attack apparatus and dump site pumper have topped off their onboard water tanks.

____

____

_________ 28. One of the most accurate ways to measure the amount of water left in the tank after dumping is to weigh the apparatus when it is completely full and then again after the tank has been dumped.

____

____

Multiple Choice

C. Write the letter of the best answer on the blank before each statement.

____ 1. Which of the following is an advantage of water shuttle operations over relay operations?

a. More reliable

b. Require fewer apparatus

c. Safer

d. Easier to pack up after incident

____ 2. Which of the following is ***not*** typically a primary factor in the choice between relay pumping operations and water shuttle operations?
   a. Tradition
   b. Cost
   c. Jurisdiction's equipment
   d. Mutual aid agencies' equipment

____ 3. Fill site pumpers should have a minimum pump capacity of ____.
   a. 750 gpm *(3 000 L/min)*
   b. 1,000 gpm *(4 000 L/min)*
   c. 1,250 gpm *(5 000 L/min)*
   d. 1,500 gpm *(6 000 L/min)*

____ 4. Light-duty-truck type fill site pumpers have capacities to ____.
   a. 1,000 gpm *(4 000 L/min)*
   b. 1,200 gpm *(4 800 L/min)*
   c. 1,400 gpm *(5 600 L/min)*
   d. 1,600 gpm *(6 400 L/min)*

____ 5. The majority of tenders today carry ____ of water.
   a. 1,000 – 2,000 gallons *(4 000 L to 8 000 L)*
   b. 1,500 – 2,500 gallons *(6 000 L to 10 000 L)*
   c. 1,500 – 3,000 gallons *(6 000 L to 12 000 L)*
   d. 2,000 – 5,000 gallons *(8 000 L to 20 000 L)*

____ 6. Vehicle weight restrictions generally limit single, rear-axle trucks to a maximum tank capacity of ____.
   a. 1,000 gallons *(4 000 L)*
   b. 1,250 gallons *(5 000 L)*
   c. 1,500 gallons *(6 000 L)*
   d. 1,750 gallons *(7 000 L)*

____ 7. Tenders that have water tanks smaller than ____ and quick unloading times are the most efficient for water shuttle operations.
   a. 1,000 gallons *(4 000 L)*
   b. 1,500 gallons *(6 000 L)*
   c. 2,000 gallons *(8 000 L)*
   d. 2,500 gallons *(10 000 L)*

____ 8. Which of the following is ***not*** a use for which a tender with a 4,000 gallon *(16 000 L)* tank would be best suited?
   a. Water shuttle
   b. Nurse tanker
   c. Refill brush fire apparatus
   d. Refill water-drop aircraft

____ 9. NFPA 1901 requires tenders equipped with a fire pump to have a tank-to-pump line capable of supplying the pump with ____ until at least ____ of the tank is emptied.
   a. 250 gpm *(950 L/min)*; 90 percent
   b. 500 gpm *(1 900 L/min)*; 80 percent
   c. 750 gpm *(2 850 L/min)*; 75 percent
   d. 1,000 gpm *(3 785 L/min)*; 60 percent

____ 10. NFPA 1901 requires tenders to have a least one external fill connection that allows the tank to be filled at a minimum rate of ____.

a. 250 gpm *(950 L/min)*
b. 500 gpm *(1 900 L/min)*
c. 750 gpm *(2 850 L/min)*
d. 1,000 gpm *(3 785 L/min)*

____ 11. NFPA 1901 requires tenders to have at least one large tank discharge that can discharge ____ percent of the tank volume at an average rate of 1,000 gpm *(3 785 L/min)*.

a. 80
b. 85
c. 90
d. 95

____ 12. Tenders that use MDH supply lines for refilling should have at least ____ external fill connection(s) piped directly to the tank.

a. One
b. Two
c. Three
d. Four

____ 13. Which of the following statements about direct tank fill inlets is true?

a. Fill inlets near the top of the tank have a significant hydraulic advantage over inlets near the bottom of the tank.
b. Most water tanks are less than 6 feet *(2 m)* tall.
c. Back pressure on a fill inlet near the bottom of a 6 foot tall *(2 m)* water tank is 6 psi *(1.0 kPa)*.
d. Fill inlets near the bottom of the tank are generally more hazardous than those near the top.

____ 14. Each tender should be equipped with at least ____ dump valves.

a. Two
b. Three
c. Four
d. Five

____ 15. Gravity dumps usually employ ____ or larger round or square piping with a valve that extends to the exterior of the apparatus.

a. 6-inch *(150-mm)*
b. 8-inch *(200-mm)*
c. 10-inch *(250-mm)*
d. 12-inch *(300-mm)*

____ 16. Who should select the closest suitable water supply source for a particular incident?

a. Fire chief
b. Driver/operator
c. Incident commander
d. Fire engineer

____ 17. Which of the following statements about selecting a fill or dump site is true?

a. Fill sites should be capable of supplying at least 1,500 gpm *(6 000 L/min)*.
b. The best sites are those that the apparatus drives through without changing direction.
c. If maneuvering is unavoidable, it is better to do it when the tank is full.
d. The confusion that multiple fill and dump sites cause outweighs their advantages.

____ 18. Which of the following statements about water shuttles in the incident command system is true?

a. Many departments find that it is most effective to view the water shuttle operation and the fireground operation as two independent operations.

b. The water supply group supervisor works independently of the incident commander.

c. When the water supply sector is established, driver/operators report directly to the incident commander.

d. Water supply groups are established for smaller incidents, water supply sectors for larger incidents.

____ 19. Which of the following statements about water shuttles in the incident command system is true?

a. The incident commander always personally monitors the water demand at the dump site.

b. The incident commander should be in close contact with the individuals in charge of fill sites and dump sites.

c. When two or more shuttles are required, it is necessary to establish a water supply branch.

d. The first arriving officer appoints individuals to head the fill and dump site operations of each branch.

____ 20. Which of the following pump panel locations affords the best view of fill or dump site operations?

a. Top

b. Intake side

c. Discharge side

d. Front

____ 21. When positioning a supply pumper at a hydrant, the driver/operator should connect, at a minimum, ____ between the hydrant and the pumper.

a. One large-diameter intake hose

b. Two medium-diameter intake hoses

c. One large- and one medium-diameter intake hose

d. Two large-diameter intake hoses

____ 22. Portable high-volume pumps may be used to relay water from the source to the fill site pumper if the distance is no more than ____.

a. 100 feet *(30 m)*

b. 200 feet *(60 m)*

c. 300 feet *(90 m)*

d. 400 feet *(120 m)*

____ 23. The most common and efficient fill connections on tenders are ____.

a. One LDH direct tank fill connection ***and*** one 2½-inch *(65-mm)* connection at the rear of the vehicle

b. One 2½-inch *(65-mm)* connection on both sides of the vehicle

c. One LDH direct tank fill connection ***or*** two 2½-inch *(65-mm)* connections at the rear of the vehicle

d. One 3½-inch connection ***and*** one 2½-inch *(65-mm)* connection at the rear of the vehicle

____ 24. If a traffic cone is used to mark the tender's stopping place, the tender driver should stop with ____ adjacent to the cone.

a. The front bumper
b. The front axle
c. The front of the cab
d. The driver's side door window

____ 25. If available, where should gate valves be placed on lines from the fill site pumper to the tender?

a. At the end of the last hose section
b. Between the last and the next-to-last hose sections
c. At the end of the second hose section
d. At the discharge outlet

____ 26. Which of the following is the most practical approach if a large in-line gate valve is not available for an operation using large diameter hose?

a. Hose clamps
b. Apparatus discharge valve
c. LDH manifold
d. Reducers

____ 27. Which of the following is the most practical approach if some tenders to be filled have 2½-inch *(65-mm)* lines and some have LDH?

a. Hose clamps
b. Apparatus discharge valve
c. LDH manifold
d. Reducers

____ 28. The jets in the fill pipe of an overhead fill system should be pumped at ____.

a. 100 psi *(700 kPa)*
b. 150 psi *(1 050 kPa)*
c. 200 psi *(1 400 kPa)*
d. 250 psi *(1 750 kPa)*

____ 29. The jets in the fill pipe of an overhead fill system should provide ____ through 4-inch *(65-mm)* pipe.

a. 700 to 800 gpm *(2 800 L/min to 3 200 L/min)*
b. 800 to 1,000 gpm *(3 200 L/min to 4 000 L/min)*
c. 1,000 to 1,200 gpm *(4 000 L/min to 4 800 L/min)*
d. 1,200 to 1,500 gpm *(4 800 L/min to 6 000 L/min)*

____ 30. Which of the following is the proper procedure for disconnecting a fill line to a tender?

a. Close intake valve at tender, disconnect hose at gate valve, signal first tender driver to leave and second to proceed cautiously into position.
b. Close gate valve, close discharge valve on fill site pumper, bleed hose behind gate valve, disconnect hose from tender.
c. Close gate valve, bleed last section of line, disconnect hose from tender.
d. Throttle back fill site pumper, close discharge valve, bleed line, disconnect hose at intake valve on tender.

____ 31. Which of the following statements about shutting down the fill site is true?
- a. The fill site pumper should shut down as soon as the incident commander determines the shuttle operation is no longer necessary.
- b. It is generally recommended that the apparatus closest to the scene be released first.
- c. Most jurisdictions prefer not to refill their tenders until they have returned to quarters.
- d. If the water supply was a static source, all pumps and equipment should be flushed.

____ 32. Which of the following is an advantage of a direct pumping operation?
- a. It works well for tenders with small capacity pumps.
- b. Tenders can connect and disconnect without interruption.
- c. The tender does not have to park directly next to the attack pumper.
- d. It easily provides a constant supply of water to the attack pumper.

____ 33. The primary advantage of nurse tender operations is that ____.
- a. They work well for tenders with small capacity pumps.
- b. The fire is often controlled before the tender's tank needs to be refilled.
- c. Tractor-trailer tenders are highly effective as shuttle apparatus.
- d. The dumping time is significantly lower than for other operations.

____ 34. Which of the following is the most efficient type of dump site operation?
- a. Direct pumping
- b. Nurse tender
- c. Lateral series
- d. Portable water tank

____ 35. The primary disadvantage of a portable water tank operation is that ____.
- a. Ensuring a constant supply of water to the attack pumper is relatively difficult.
- b. Multiple tank operations require a large working area.
- c. Shuttle tenders must be equipped with fire pumps.
- d. The dumping time is significantly higher than for other operations.

____ 36. A gravity dump's discharge pressure when the water column in the tank is 3 feet *(0.9 m)* is ____.
- a. 0.434 psi *(3 kPa)*
- b. 0.868 psi *(6 kPa)*
- c. 1.302 psi *(9 kPa)*
- d. 1.736 psi *(12 kPa)*

____ 37. The maximum flow rate for which single portable tank operations work is ____.
- a. 300 gpm *(1 200 L/min)*
- b. 500 gpm *(2 000 L/min)*
- c. 600 gpm *(2 400 L/min)*
- d. 750 gpm *(3 000 L/min)*

____ 38. The most common portable water tank is the ____ type.
- a. Folding
- b. Sectional
- c. Self-supporting
- d. Rigid

____ 39. A portable water tank should have a capacity at least ____ the capacity of the water tank on the apparatus carrying it.

a. 500 gallons*(2 000 L)* less than
b. Equal to
c. 500 gallons*(2 000 L)* more than
d. 1,000 gallons *(4 000 L)* more than

____ 40. Which of the following strainer types should be used for drafting from a portable tank?

a. Barrel-type
b. Sinker
c. Floating
d. Low-level

____ 41. What should be done to insure that the dump site pumper does not lose prime when discharge lines are shut down?

a. The dump site pumper should throttle back.
b. The dump site pumper should flow a booster line back into the portable tank.
c. The tender should flow a booster line into the portable tank.
d. The tender should flow a booster line directly to the pumper.

____ 42. The most common multiple portable tank operations in most jurisdictions range use from ____ portable tanks.

a. Two to five
b. Two to seven
c. Three to eight
d. Five to ten

____ 43. The basic goal of multiple tank operations is ____.

a. Keep the first tank in the chain full at all times.
b. Keep all tanks in the operation at a minimum of one half their capacity.
c. Keep the number of empty tanks below one third the total number of tanks.
d. Keep the last tank in the chain full at all times.

____ 44. Simply connecting tanks by their drain openings limits an operation to no more than ____ tanks.

a. Two
b. Three
c. Four
d. Five

____ 45. What should be done if all of the portable tanks become empty at the same time during an operation?

a. Continue pumping from the dump site pumper's onboard tank and prepare firefighters to withdraw.
b. Suspend the fire attack until tank is full enough to draft from.
c. Continue pumping from the dump site pumper's onboard tank and withdraw firefighters from hazardous positions.
d. Suspend the fire attack until at least one tank is full and a second is full enough to draft from.

____ 46. If a 2,500 gallon *(10,000 L)* tender can make ¾-mile *(1.2 km)* round trip from the dump site to the fill site and back in 9 minutes, what is its flow rate?

a. 278 gpm *(1 100 L/min)*
b. 208 gpm *(832 L/min)*
c. 217 gpm *(868 L/min)*
d. 250 gpm *(1 000 L/min)*

____ 47. The ISO formulas for evaluating tender performance by dividing the shuttle operation into travel time and handling time are used to rate ____.

a. Industrial fire brigades
b. Rural fire departments
c. Urban fire departments
d. Military fire fighting techniques

____ 48. ISO allows ____ percent of the tender's total tank capacity to be used when calculating its flow rate.

a. 75
b. 80
c. 85
d. 90

____ 49. Using the ISO formulas, calculate the flow rate for a 3,000 gallon tanker that will be shuttling water 2.5 miles in each direction. Assume that fill and dump rates are 1,000 gpm and that maneuvering and make-and-break times at each site total 2 minutes.

a. 181 gpm
b. 211 gpm
c. 270 gpm
d. 292 gpm

____ 50. Using the ISO formulas, calculate the flow rate for a 9,000 liter tanker that will be shuttling water 7 kilometers in each direction. Assume that fill and dump rates are 4 000 L/min and that maneuvering and make-and-break times at each site total 2 minutes.

a. 415 L/min
b. 489 L/min
c. 512 L/min
d. 663 L/min

Chapter 15 Foam Equipment and Systems

Matching

A. Match to their definitions terms associated with foam equipment and systems. Write the appropriate letters on the blanks.

____ 1. Type of foam that must be proportioned and aerated before it can be used

____ 2. Raw foam liquid before the introduction of air and water

____ 3. Device that introduces foam concentrate into the water stream

____ 4. Mixture of foam concentrate and water before the introduction of air

____ 5. Completed product after air is introduced into foam solution

____ 6. Petroleum based Class B fuel

____ 7. Flammable liquid that is miscible in water

____ 8. Creates a barrier between the fuel and the fire

____ 9. Prevents the release of flammable vapors

____ 10. Mixing water with foam concentrate to form a foam solution

____ 11. The process of foam's bubbles breaking down and releasing water

____ 12. Increases a foam solution's volume when it is aerated

____ 13. Suction effect caused by low pressure area near the outlet side of an eductor

____ 14. Pressure at the outlet of an eductor

a. Drainage
b. Back pressure
c. Foam proportioner
d. Chemical
e. Venturi principle
f. Mechanical
g. Foam
h. Proportioning
i. Expansion
j. Polar solvent
k. Foam concentrate
l. Foam solution
m. Suppress
n. Hydrocarbon
o. Separate

True/False

B. Write *True* or *False* before each of the following statements. Correct those statements that are false.

_______ 1. Chemical foams are obsolete.

_______ 2. Gasoline with 10 percent or more solvent additives should be considered a polar solvent and handled as such during an emergency.

_______ 3. Some modern alcohol-resistant concentrations are proportioned as high as ten parts of water.

_______ 4. Class A foams can be adjusted to a lower percentage to produce a thin foam that rapidly sinks into a fuel's surface.

_______ 5. The application rate for Class A foam is the same as the minimum critical flow rate for water.

_______ 6. A long drain time means rapid wetting.

_______ 7. Besides extinguishing fires, Class B foam can be used to suppress vapors from unignited spills.

_________ 8. The application rate for protein and fluoroprotein foams on nondiked hydrocarbon fuel spill fires using portable extinguishing equipment is 0.10 gpm/ft^2 (4.1 [L/min]/m^2) for 15 minutes.

_________ 9. Unignited spills require the same application rates as ignited spills.

_________ 10. Concentrates designed to be used at 3% on hydrocarbon fuels and at 6% on polar solvents are commonly referred to as thirty-six concentrates.

_________ 11. Alcohol-resistant AFFF should be sprayed over the top of the fuel rather than plunged into it.

_________ 12. For the nozzle and in-line eductor to operate properly, both must have the same flow rate.

_________ 13. Installed in-line eductor systems are effective for proportioning Class A foams.

_________ 14. An around-the-pump foam proportioner does not allow plain water and foam to be discharged from the pump at the same time.

______ 15. In a variable-flow demand-type balanced pressure proportioner, a foam concentrate pump supplies foam concentrate to a venturi-type proportioning device built into the water line.

______ 16. CAFS proportioners typically use foam eductors.

______ 17. Most structural and wildland fire attacks using CAFS are done with an air flow rate of 0.5 to 1.0 cfm *(0.015 m^3/min to 0.03 m^3/min).*

______ 18. The discharge orifice of a solid bore nozzle with a Class A compressed-air foam system should be no greater than one half the diameter of the hose.

______ 19. Air-aspirating foam nozzles may be used with Class A foams in wildland applications.

______ 20. Standard fixed-flow or automatic master stream nozzles may be used to deliver foam.

______ 21. The roll-on foam application technique is suitable for all types of flammable and combustible liquid fires.

______ 22. The rain-down foam application method should be used only when the other methods are not feasible.

Multiple Choice

C. Write the letter of the best answer on the blank before each statement.

____ 1. In most cases, who is responsible for the proper assembly and operation of foam fire fighting equipment?
   a. Firefighter II
   b. Driver/operator
   c. Company officer
   d. Fire engineer

____ 2. Most fire fighting foam concentrates are intended to be mixed with ____ percent water.
   a. 94 – 99.9
   b. 90 – 97
   c. 90 – 99.9
   d. 70 – 90

____ 3. Which of the following proportioning methods works by passing a stream of water through a device with a restricted diameter?
   a. Injection
   b. Batch mixing
   c. Induction
   d. Premixing

____ 4. Which of the following proportioning methods uses an external pump or head pressure to force foam concentrate into the fire stream?
   a. Injection
   b. Batch mixing
   c. Induction
   d. Premixing

____ 5. In which of the following proportioning methods is an appropriate amount of foam concentrate poured directly into a tank of water?
   a. Injection
   b. Batch mixing
   c. Induction
   d. Premixing

____ 6. Batch mixing ____.
   a. Is especially useful with Class B foams
   b. Provides a continuous supply of foam for large incidents
   c. Allows for accurate proportioning
   d. Eliminates time needed for thorough mixing

____ 7. In most instances, premixed solutions are discharged from ____.
   a. A pump and non-pressure-rated atmospheric storage tank
   b. An eductor
   c. A venturi
   d. A pressure-rated tank

____ 8. Which of the following is not a common method of storing foam concentrate in the municipal and wildland fire services?

a. Five-gallon *(20 L)* plastic pails

b. 20-gallon *(80 L)* galvanized steel cans

c. 55-gallon *(250 L)* plastic barrels

d. Apparatus tanks

____ 9. Which of the following statements about foam concentrate storage is true?

a. Smaller foam concentrate tanks are located directly above the fire pump area.

b. Designs incorporating foam concentrate tanks as an integral cell within the apparatus are outmoded.

c. Foam concentrate tanks on municipal pumpers range from 100 to 400 gallons *(400 L to 1 600 L)*.

d. Standard vented atmospheric storage tanks are acceptable for foam concentrate.

____ 10. Which of the following statements about Class A foam is true?

a. It increases surface tension.

b. It should not be used with solid stream nozzles.

c. Its corrosive and supercleaning effects are harmful to fire apparatus components.

d. Under ordinary fire suppression conditions, harm to the environment should not be a concern.

____ 11. Class A foam concentrates are mixed in proportions of ____ percent.

a. 1.0 to 3.0

b. 0.1 to 0.3

c. 0.3 to 1.0

d. 0.1 to 3.0

____ 12. Foam concentrations higher than ____ percent do not appear to improve fire fighting performance when used with standard fog nozzles.

a. 0.1

b. 0.5

c. 1.0

d. 3.0

____ 13. Which of the following is a correct guideline for Class A foam proportioning?

a. Fire attack and overhaul with standard fog nozzles; 0.2 to 3.0 percent foam concentrate

b. Exposure protection with standard fog nozzles; 1.0 to 1.5 percent foam concentrate

c. Any application with air-aspirating foam nozzles; 0.3 to 0.7 percent foam concentrate

d. Any application with compressed air foam systems; 2.0 to 2.5 percent foam concentrate

____ 14. Which of the following foams is most desirable for vertical surfaces?

a. Wet
b. Dry
c. Heavy
d. Medium

____ 15. Which of the following statements about Class B foams is true?

a. Protein-based foams are derived from vegetable oils.
b. Protein-based foams generally have a longer shelf-life than synthetic foams.
c. Foams made to United States military specifications by different manufacturers can be mixed as long as they are of the same type.
d. Synthetic foam is made from enzymes.

____ 16. Today's Class B foams are mixed in proportions from ____ percent.

a. 1 to 6
b. 3 to 6
c. 3 to 10
d. 6 to 10

____ 17. Medium expansion foams are typically used at either ____ percent concentrations.

a. ½, 1, or 1½
b. 1, 1½, or 2
c. 1½, 2, or 3
d. 2, 2½, or 3

____ 18. What is the key factor in choosing a foam concentrate for a specific application?

a. Proportioning
b. Method of aspiration
c. Contents
d. Expansion

____ 19. Medium expansion foam is most commonly used at a rate ____.

a. Up to 20:1
b. From 20:1 to 100:1
c. From 20:1 to 200:1
d. From 200:1 to 1,000:1

____ 20. What NFPA standard includes tables that list foam application rate requirements for ignited fuels?

a. 1001
b. 11
c. 1901
d. 1170

____ 21. Combustible liquid hydrocarbon fuel fires in fixed roof storage tanks using portable extinguishing equipment require a ____ application.

a. 45-minute
b. 50-minute
c. 60-minute
d. 75-minute

____ 22. Which of the following foam concentrate types is derived from naturally occurring sources of protein and rarely used in the fire service today?

a. Aqueous film forming foam
b. Fluoroprotein foam
c. Regular protein foam
d. Film forming fluoroprotein foam

____ 23. What combination protein-based and synthetic-based foam provides a strong security blanket for long-term vapor suppression?
   a. Aqueous film forming foam
   b. Fluoroprotein foam
   c. Regular protein foam
   d. Film forming fluoroprotein foam

____ 24. Which of the following foam concentrate types combines AFFF's fast fire knockdown capability with fluoroprotein foam's long-lasting heat resistance?
   a. Aqueous film forming foam
   b. Fluoroprotein foam
   c. Regular protein foam
   d. Film forming fluoroprotein foam

____ 25. Which of the following is the most commonly used foam concentrate type in the fire service today?
   a. Aqueous film forming foam
   b. Fluoroprotein foam
   c. Regular protein foam
   d. Film forming fluoroprotein foam

____ 26. Which of the following is ***not*** a basic application for high-expansion foams?
   a. Concealed spaces
   b. Fixed extinguishing systems
   c. Class A fires
   d. Class B fires

____ 27. For medium expansion uses, high-expansion foam concentrates have expansion ratios ranging from ____.
   a. 10:1 to 200:1
   b. 20:1 to 100:1
   c. 20:1 to 200:1
   d. 200:1 to 1,000:1

____ 28. In general, foam proportioning devices operate by one of two basic principles, ____.
   a. Induction and gravitation
   b. Gravitation and inertia
   c. Injection and inertia
   d. Induction and injection

____ 29. What type of foam proportioner is designed to be either directly attached to the pump panel discharge or connected at some point in the hose lay?
   a. Self-educting master stream nozzle
   b. In-line foam eductor
   c. Foam nozzle eductor
   d. High-flow needle eductor

____ 30. Pressure at the outlet of an in-line eductor must not exceed ____ percent of the eductor inlet pressure.

a. 55 to 60
b. 60 to 65
c. 65 to 70
d. 70 to 75

____ 31. The rated inlet pressure of an in-line eductor is usually ____.

a. 50 to 100 psi *(350 kPa to 700 kPa)*
b. 100 to 150 psi *(700 kPa to 1 050 kPa)*
c. 150 to 200 psi *(1 050 kPa to 1 400 kPa)*
d. 200 to 250 psi *(1 400 kPa to 1 750 kPa)*

____ 32. The foam concentrate inlet to an in-line eductor should not be more than ____ above the liquid surface of the foam concentrate.

a. 4.5 feet *(1.5 m)*
b. 6 feet *(2 m)*
c. 7.5 feet *(2.5 m)*
d. 9 feet *(3 m)*

____ 33. What type of proportioning device compromises firefighter safety because the foam concentrate needs to be moved with the nozzle?

a. Self-educting master stream nozzle
b. In-line foam eductor
c. Foam nozzle eductor
d. High-flow needle eductor

____ 34. What type of foam eductor uses a venturi pickup tube in the center bore of the nozzle?

a. Self-educting master stream nozzle
b. In-line foam eductor
c. Foam nozzle eductor
d. High-flow needle eductor

____ 35. Self-educting master stream nozzles are used when flows greater than ____ are required.

a. 250 gpm *(1 750 kPa)*
b. 300 gpm *(2 100 kPa)*
c. 350 gpm *(2 450 kPa)*
d. 400 gpm *(2 800 kPa)*

____ 36. What type of inline eductor allows the foam concentrate supply to be as far away as 3,000 feet *(900 m)* from a self-educting master stream foam nozzle?

a. High-flow needle eductor
b. Pressure enhanced valve
c. Rotary gain pickup
d. Jet ratio controller

____ 37. A jet ratio controller proportions concentrate as a ____ percent solution.

a. 3
b. 6
c. 66½
d. 77½

____ 38. What type of apparatus-mounted foam proportioning system is a bypass proportioner?

a. Variable-flow variable-rate direct injection
b. Installed in-line eductor
c. Bypass-type balanced pressure proportioner
d. Around-the-pump

____ 39. What type of apparatus-mounted foam proportioning system uses a small return line from the discharge side of the pump back to the intake side?

a. Variable-flow variable-rate direct injection
b. Installed in-line eductor
c. Bypass-type balanced pressure proportioner
d. Around-the-pump

____ 40. What type of apparatus-mounted foam proportioning system has a foam concentrate line connected to each fire pump discharge outlet?

a. Variable-flow variable-rate direct injection
b. Installed in-line eductor
c. Bypass-type balanced pressure proportioner
d. Around-the-pump

____ 41. Which of the following statements about bypass-type balanced pressure proportioners is true?

a. The orifice of the foam concentrate line is adjustable at the point where it connects to the discharge line.
b. They are most commonly used on municipal pumpers.
c. They can supply either plain water or foam concentrate, but not both at the same time.
d. They are not limited by the need for a foam pump with a PTO.

____ 42. What type of apparatus-mounted proportioning system requires monitoring of the water flow and adjusting the speed of a positive displacement foam concentrate pump to provide the correct amount of foam concentrate injection?

a. Variable-flow variable-rate direct injection
b. Installed in-line eductor
c. Bypass-type balanced pressure proportioner
d. Around-the-pump

____ 43. Which of the following statements about variable-flow variable-rate direct injection systems is true?

a. They provide foam concentrate rates from 3 to 6 percent.
b. They can be used with all Class A and Class B foam concentrates.
c. The injection point can be in the piping beyond manifolds or distribution to multiple fire pump discharges.
d. The system automatically adjusts to changes in water flow when nozzles are opened or closed.

____ 44. Which of the following is ***not*** an advantage of variable-flow demand-type proportioners?

a. Foam concentrate flow and pressure match system demand.

b. Pressure drops across the discharge are lower than those on standard pumpers.

c. There is no recirculation back to the foam concentrate tank.

d. Water and foam can be discharged simultaneously from any combination of outlets up to rated capacity.

____ 45. How much foam concentrate is needed to batch mix a 6 percent solution in a 750 gallon *(3 000 L)* tank?

a. 45 gallons *(180 L)*

b. 22.5 gallons *(90 L)*

c. 30 gallons *(120 L)*

d. 60 gallons *(240 L)*

____ 46. Which of the following statements about batch mixing is true?

a. In general it is only used with alcohol-resistant AFFF concentrates.

b. Class A foam solutions retain their foaming properties up to 48 hours.

c. Foam solutions are excellent lubricating agents.

d. It does not allow continuous foam discharge on large incidents.

____ 47. What type of foam system introduces compressed air into the foam solution prior to discharge into the hoseline?

a. Variable-flow demand

b. High-energy

c. Pressurized batch mix

d. Variable-flow variable-rate direct injection

____ 48. What type of high-energy foam generating system uses a standard centrifugal fire pump to supply the water and a direct-injection foam proportioning system attached to the discharge side of the fire pump?

a. WES

b. WEPS

c. SMART

d. CAFS

____ 49. Which of the following is a disadvantage of a compressed air foam system?

a. The air bubbles are not very durable.

b. CAFS-produced foam does not adhere to the fuel surface as long as low-energy foam.

c. Compressed air accentuates hose reaction in the event of a hose rupture.

d. Hoselines containing high-energy foam are heavier than those containing low-energy foam.

____ 50. Low-energy foam solutions, aerate the foam in the ____.

a. Nozzle

b. Pump

c. Eductor

d. Hoseline

____ 51. IFSTA defines a handline nozzle as "any nozzle that one to three firefighters can safely handle and that flows less than ____.

a. 200 gpm *(800 L/min)*
b. 250 gpm *(1 000 L/min)*
c. 300 gpm *(1 200 L/min)*
d. 350 gpm *(1 400 L/min)*

____ 52. The use of solid bore nozzles with foam is limited to ____ applications.

a. All Class A
b. Class A CAFS only
c. All Class B
d. Class B CAFS only

____ 53. Either fixed-flow or automatic flow fog nozzles produce a ____ foam.

a. High-expansion, long-lasting
b. High-expansion, short-lasting
c. Low-expansion, long-lasting
d. Low-expansion, short-lasting

____ 54. A fog nozzle's best application with foam is with ____ foams

a. Alcohol-resistant AFFF and Class A
b. FFFP and Class B
c. Regular AFFF and Class A
d. Regular AFFF and FFFP

____ 55. What type of nozzle inducts air into the foam solution by a venturi action?

a. Fixed-flow fog
b. Automatic fog
c. Air-aspirating
d. Solid bore

____ 56. The ratio of air to foam in medium expansion foam generators ranges from ____.

a. 20:1 to 100:1
b. 20:1 to 200:1
c. 200:1 to 500:1
d. 200:1 to 1,000:1

____ 57. The major difference between a water-aspirating foam generator and a mechanical blower is that ____.

a. The water-aspirating generator does not have a powered fan and the mechanical blower does.
b. The water-aspirating foam generator should not be used with Class A foam.
c. The mechanical blower generator is limited to medium expansion foam.
d. The water aspirating generator produces a higher-air-volume foam than the mechanical blower.

____ 58. Which of the following is ***not*** a common reason for failure to generate foam or for generating poor quality foam?

a. Failure to match eductor and nozzle flow
b. Improper cleaning of proportioning equipment
c. Nozzle too far above eductor
d. Too long a hose lay on intake side of the eductor

____ 59. What foam application method directs the foam stream on the ground near the front edge of a burning liquid pool?

a. Rain-down
b. Roll-on
c. Cross-spray
d. Bank-down

____ 60. What foam application method directs the foam stream off an elevated object near a burning pool of liquid or unignited liquid spill?

a. Rain-down
b. Roll-on
c. Cross-spray
d. Bank-down

____ 61. What foam application method directs the stream into the air above the fire or spill and allows the foam to float gently down onto the surface of the fuel?

a. Rain-down
b. Roll-on
c. Cross-spray
d. Bank-down

Identify

D. Identify the following abbreviations associated with Foam Equipment and Systems. Write the correct interpretation before each.

________________________________ 1. CAFS

________________________________ 2. FFFP

________________________________ 3. AFFF

________________________________ 4. JRC

________________________________ 5. WES

________________________________ 6. WEPS

Chapter 16 Apparatus Testing

Matching

A. Match to their definitions terms associated with apparatus testing. Write the appropriate letters on the blanks. Definitions continued and terms repeated on the next page.

____ 1. Includes manufacturer's tests, pump certification tests, and acceptance tests

____ 2. Determines whether the pump and pump piping can withstand pressures normally encountered during fire fighting operations

____ 3. Determines whether the fire pump system operates as intended after it is installed on the apparatus chassis

____ 4. Conducted to assure the purchaser that the apparatus meets bid specifications

____ 5. The total work done to get water into, through, and out of the pump

6. Performed to check priming device, pump, and hard intake hose for air leaks

____ 7. Checks the overall condition of the engine and the pump

____ 8. Helps to ensure that a safe level of pressure is maintained when valves are closed at a variety of discharge pressures

____ 9. Measures the accuracy of foam proportioning equipment by checking the volume of foam concentrate drawn through the system while it is in operation

____ 10. Collects the discharge from the foam concentrate pump in a calibrated container for a specified time

____ 11. Measures the amount of foam concentrate in foam solution

a. Pressure control test
b. Conductivity
c. Net pump discharge pressure
d. Foam concentrate pump discharge volume method
e. Preservice test
f. Service test
g. Refractometer
h. Vacuum test
i. Pump certification test
j. Foam solution conductivity testing
k. Hydrostatic test
l. Foam concentrate displacement method
m. Acceptance test
n. Pumping test

____ 12. Checks the quality of synthetic-based foams produced by foam proportioning equipment and systems

____ 13. The ability of a substance to conduct an electrical current

a. Pressure control test
b. Conductivity
c. Net pump discharge pressure
d. Foam concentrate pump discharge volume method
e. Preservice test
f. Service test
g. Refractometer
h. Vacuum test
i. Pump certification test
j. Foam solution conductivity testing
k. Hydrostatic test
l. Foam concentrate displacement method
m. Acceptance test
n. Pumping test

True/False

B. Write *True* or *False* before each of the following statements. Correct those statements that are false.

_______ 1. Fire apparatus tests affect insurance rates in the jurisdiction.

____

____

_______ 2. The driver/operator often performs service tests or at least assists the mechanics who perform them.

____

____

_______ 3. Bid specifications for fire apparatus should contain a clause that requires them to meet the pertinent chapters of NFPA 1901 or 1906.

____

____

_________ 4. Fire department personnel generally are involved in performing manufacturer's certification tests.

_________ 5. Road tests should be conducted with the vehicle's water and foam tanks, hose beds, and equipment bays unloaded.

_________ 6. The acceptance test should include another pump test, even if a certification test was performed at the factory.

_________ 7. Most jurisdictions prefer to pump from a hydrant during pumper service tests.

_________ 8. The gauge used to check pump discharge pressure during service tests should have a range of 30 in.Hg *(100 kPa)* to zero if it is a vacuum gauge or 30 in.Hg vacuum to 150 psi *(1 050 kPa)* if it is a compound gauge.

_________ 9. NFPA 1911 prohibits using flowmeters instead of a pitot gauges for service tests.

_________ 10. An ideal time to test hard intake hose is following the dry vacuum test.

_________ 11. If discharge pressure gauges and flowmeters are not working properly, the driver/operator might supply dangerously insufficient or excessive amounts of water to firefighters operating hose streams.

____

____

_________ 12. Flowmeter tests require that all discharges be checked simultaneously.

____

____

_________ 13. Tank-to-pump flow tests should be performed only for the NFPA minimum regardless of the flow capacity specified when the apparatus was ordered.

____

____

_________ 14. If a fire pump tests lower than 90 percent of its capabilities when it was new and underwent certification testing, it may be kept in service and given a lower rating based on its actual performance.

____

____

_________ 15. The readout on a direct reading conductivity meter may not indicate the actual percentage of foam concentrate in the solution.

____

____

Multiple Choice

C. Write the letter of the best answer on the blank before each statement.

____ 1. Once it is placed into service, fire apparatus must be tested at least every ____ months.

a. 6 c. 24

b. 12 d. 36

____ 2. Which NFPA standard requires that driver/operators be able to perform functional tests on foam systems?

a. 1001 c. 1901

b. 1002 d. 2001

____ 3. NFPA ____ is the Standard for Automotive Fire Apparatus.
a. 1901
b. 1902
c. 1905
d. 1906

____ 4. NFPA ____ is the Standard for Wildland Fire Apparatus.
a. 1901
b. 1902
c. 1905
d. 1906

____ 5. NFPA requires that fire apparatus must accelerate to 35 mph *(56 km/h)* from a standing start within ____ seconds.
a. 20
b. 25
c. 30
d. 35

____ 6. NFPA requires that fire apparatus must come to a full stop from 20 mph *(32 km/h)* within ____.
a. 25 feet *(7.6 m)*
b. 30 feet *(9.2 m)*
c. 35 feet *(10.7m)*
d. 50 feet *(15.3 m)*

____ 7. Many departments prefer to write performance-based specifications rather than engineering/equipment specifications because of ____.
a. The cost savings
b. Their special requirements
c. The insurance benefits
d. Their greater flexibility

____ 8. Hydrostatic tests test pumps at ____ for three minutes.
a. 100 psi *(1 690 kPa)*
b. 150 psi *(1035 kPa)*
c. 250 psi *(1 725 kPa)*
d. 350 psi *(2 415 kPa)*

____ 9. Who performs pump certification tests?
a. The manufacturer
b. The purchasing department
c. The fire marshal
d. An independent testing organization

____ 10. Apparatus equipped with a fire pump with a rated capacity less than 750 gpm *(3 000 L/min)* do not have to undergo the ____ test.
a. Pumping engine overload
b. Pressure control system
c. Vacuum
d. Water tank-to-pump

____ 11. If an apparatus fails to perform according to the requirements of the bid specifications, it should be ____.
a. Provisionally accepted
b. Accepted with satisfactory alternative "credits"
c. Rejected
d. Accepted with an appropriate cost adjustment

____ 12. At altitudes above 2,000 feet *(610 m)* the ____ test must be performed during acceptance testing.

a. Pumping engine overload
b. Pressure control system
c. Vacuum
d. Water tank-to-pump

____ 13. Which of the following NFPA standards contains the requirements for fire pump service testing?

a. 1901
b. 1906
c. 1911
d. 2001

____ 14. When drafting from a static water source during pumper service tests, the water must be at least ____ deep.

a. 3 feet *(0.9 m)*
b. 4 feet *(1.2 m)*
c. 5 feet *(1.5 m)*
d. 6 feet *(1.8 m)*

____ 15. Which of the following statements about pumper service tests is true?

a. Atmospheric air temperature during testing should be between 32°F and 100° F *(0° C and 37.8° C)*.
b. The water being used for testing should be between 35°F and 209°F *(1.7° C and 98.3° C)*.
c. A 1-inch *(25-mm)* drop in barometric pressure reduces a pumper's maximum static lift by about 3 feet *(0.9 m)*.
d. Barometric pressure should be at least 29 in.Hg *(98.2 kPa)*.

____ 16. What size intake hose should be used during pumper service tests if a pumper's rated capacity is 750 gpm *(2 850 L/min)*?

a. 3 inch *(76 mm)*
b. 4 inch *(100 mm)*
c. 4½ inch *(133 mm)*
d. 5 inch *(125 mm)*

____ 17. What maximum lift is required during pumper service tests if a pumper's rated capacity is 1,500 gpm *(5 678 L/min)*?

a. 10 feet *(3 m)*
b. 15 feet *(4.5 m)*
c. 20 feet *(6 m)*
d. 25 feet *(7.5 m)*

____ 18. The minimum size hose that may be used for discharging water during pumper service tests is ____.

a. 1½ inch *(38 mm)*
b. 2 inch *(50 mm)*
c. 2½ inch *(65 mm)*
d. 3 inch *(76 mm)*

____ 19. Pumper service tests for a 1,000 gpm rated pump should use ____ line(s) with a 2-inch nozzle.

a. One
b. One or two
c. Two
d. Two or three

____ 20. What size nozzles should be used to service test a pump with a 6 000 L/min rated pump?

a. Three 50 mm
b. Three 50 mm and one 38 mm
c. Two 45 mm and one 38 mm
d. Two 50 mm

____ 21. Calculate the pressure correction if a 1,250 gpm pump is being tested with a lift of 7 feet through 20 feet of 6-inch intake hose.

a. 12.2
b. 5.3
c. 3.7
d. 3.2

____ 22. Calculate the pressure correction if a 2 800 L/min pump is being tested with a lift of 1.9 meters through 6 meters of 115-mm intake hose.

a. 49 kPa
b. 4.1 kPa
c. 490 kPa
d. 41 kPa

____ 23. A pitot tube used for service tests should be rated at least from 0 psi *(0 kPa)* to ____.

a. 130 psi *(890 kPa)*
b. 140 psi *(965 kPa)*
c. 150 psi *(1 035 kPa)*
d. 160 psi *(1 110 kPa)*

____ 24. During service tests, all personnel should wear hearing protection if exposed to noise in excess of ____.

a. 70 dB
b. 80 dB
c. 90 dB
d. 100 dB

____ 25. If an apparatus cannot reach ____ of mercury during a vacuum test, it should be removed from service and repaired immediately.

a. 20 inches *(510 mm)*
b. 22 inches *(560 mm)*
c. 24 inches *(610 mm)*
d. 26 inches *(660 mm)*

____ 26. Any discharge pressure gauge that is off by more than ____ should be recalibrated, repaired, or replaced.

a. 5 psi *(70 kPa)*
b. 10 psi *(140 kPa)*
c. 15 psi *(210 kPa)*
d. 20 psi *(280 kPa)*

____ 27. The minimum flow rate needed to achieve valid flowmeter test results for a 2½-inch *(65-mm)* discharge pipe is ____.

a. 128 gpm *(454 L/min)*
b. 180 gpm *(682 L/min)*
c. 300 gpm *(1 135 L/min)*
d. 700 gpm *(2 650 L/min)*

____ 28. If a flow meter is off by more than ____ percent, it must be recalibrated, replaced, or repaired.

a. Five
b. Six
c. Eight
d. Ten

____ 29. Which of the following is ***not*** a probable cause of problems during service testing?

a. Lift higher than 10 feet *(3 m)*

b. High gear lockup not functioning (automatic transmission)

c. Intake hose too large

d. Excessive air leaks at intake side of pump

____ 30. Which of the following statements about foam proportioning equipment testing is true?

a. NFPA 1901 requires foam systems to be tested annually.

b. IFSTA recommends testing it semi-annually.

c. NFPA 1002 requires a Firefighter II to be able to perform this testing.

d. One testing method is to test the foam consumption rate in proportion to a known flow of water through the system.

____ 31. Which of the following statements about foam solution refractivity testing is true?

a. The first step in the testing process is to develop a base calibration curve.

b. The refractometer reflects the actual foam concentrate proportioning percentage.

c. When testing proportioning equipment that can be operated at more than one setting, only the highest setting must be tested.

d. It is recommended for protein-, fluoroprotein-, and synthetic-based foams.

____ 32. What foam proportioning equipment testing method is used when a conductivity meter that reads out in units of microsiemens per centimeter is available?

a. Direct reading conductivity testing

b. Conductivity comparison testing

c. Foam concentrate displacement method

d. Conductivity calibration curve testing

____ 33. What foam proportioning equipment testing method uses a handheld, temperature-compensated conductivity meter?

a. Direct reading conductivity testing

b. Conductivity comparison testing

c. Foam concentrate displacement method

d. Conductivity calibration curve testing

Distinguish

D. Distinguish between equipment that is needed for service tests and equipment that is recommended but not required. Write an N beside each piece of equipment below that is needed and an R beside each piece of equipment that is recommended but not required.

____ 1. Gauge to check pump intake pressure

____ 2. Clamp to hold pitot tube to test nozzle

____ 3. Test stand for gauges

____ 4. Pitot tube with knife edge

____ 5. Thermometer

____ 6. Rope, chain or test stand for securing nozzles

____ 7. Revolution counter or hand tachometer

____ 8. Stopwatch or watch with sweep second hand

Chapter 1 Answers

True/False

A.

1. True *(3)*
2. False. Driver operators *will be constantly required* to understand the written word and to convey information completely and accurately in writing. *(3, 4)*
3. True *(4)*
4. True *(6)*
5. False. Some *states and provinces require* a fire apparatus driver/operator to obtain a CDL for driving a large truck; *others do not.* *(6)*

Multiple Choice

B.

1. c *(3)*
2. b *(3)*
3. d *(4, 5)*
4. d *(3–5)*
5. a *(5)*
6. b *(5, 6)*

Identify

C.

1. National Fire Protection Association *(3)*
2. Mobile data terminal *(4)*
3. Standard operating procedure *(5)*
4. (United States) Department of Transportation *(6)*
5. Transport Canada *(6)*
6. Commercial driver's license *(6)*

Chapter 2 Answers

Matching

A.

1. m *(10)*
2. g *(11)*
3. e *(12)*
4. k *(12)*
5. f *(12)*
6. i *(12)*
7. d *(12)*
8. a *(13)*
9. l *(14)*
10. j *(17)*
11. b *(20)*
12. h *(20)*

True/False

B.

1. True *(7)*
2. True *(8)*
3. False. Most large-scale industrial foam pumpers use some form of *balanced pressure* foam proportioning system because of its reliability at large flows. *(9)*
4. False. Elevating water devices *can* be used to apply fire streams to the lower floors of a building. *(10)*
5. True *(10)*
6. False. The two proper methods for making a moving fire attack are to have firefighters using short sections of attack hose and walk alongside the apparatus *or discharge water from a nozzle that is remotely controlled from inside the cab. NFPA 1500 strictly prohibits* riding on the outside of the apparatus and discharging water as the vehicle is driven. *(12, 13)*
7. False. Fire boats have been built to deliver as much as 26,000 gpm *(98 420 L/min);* individual master stream turrets that discharge *2,000 to 3,000 gpm (8 000 L/min to 11 500 L/min)* are common. *(15)*
8. True *(15)*
9. True *(17)*
10. False. Inverters on pumpers are most commonly used to power vehicle-mounted floodlights. *(17)*
11. True *(17)*
12. True *(18)*
13. False. The most commonly used extrication tools are *hydraulically* powered. *(19)*

Multiple Choice

C.

1. c *(8)*
2. b *(8)*
3. c *(8)*
4. d *(8)*
5. a *(8, 9)*
6. c *(9)*
7. a *(9)*
8. b *(9)*
9. c *(10, 11)*
10. c *(10)*
11. c *(11)*
12. b *(11)*
13. a *(11)*
14. b *(12)*
15. d *(14)*
16. a *(14)*
17. d *(14)*
18. d *(14)*
19. c *(15)*
20. b *(16)*
21. c *(16)*
22. b *(16)*
23. d *(17)*
24. b *(17)*
25. d *(18)*
26. c *(18)*
27. c *(18, 19)*
28. b *(19)*
29. a *(19)*
30. c *(20)*
31. a *(20)*

Identify

D.

1. Gross vehicle weight *(10)*
2. Power take-off *(12)*
3. Aircraft rescue and fire fighting *(14)*
4. Crash, fire, rescue vehicle *(14)*
5. International Civil Aviation Organization *(14)*
6. (United States) Federal Aviation Administration *(14)*
7. Rapid intervention vehicle *(14)*
8. Incident command system *(20)*
9. National Wildfire Coordinating Group *(20)*

Chapter 3 Answers

Matching

A.

1. b *(22)*
2. d *(22)*
3. f *(33)*
4. e *(33)*
5. a *(33)*

True/False

B.

1. True *(22)*
2. False. *Newer* apparatus *require gentler cleaning procedures* than do older apparatus to avoid damage to new paint, detailing, and clear-coat protectants. *(24)*
3. True *(25)*
4. False. Reviewing the records of previous inspections allows the current inspector *to pay extra attention for reoccurrence of earlier problems. (26)*
5. True *(27)*
6. True *(27)*
7. False. The lenses for all lights should be in place and *not cracked or broken. (28)*
8. False. *Do not* test the operation of audible warning devices with someone standing in front of the vehicle. *(28)*
9. False. Pump intake valves at the front of the vehicle should be fully *closed when not in use. (28)*
10. False. The driver/operator may choose to make equipment inspections *during side inspections. (29, 30)*
11. True *(34)*
12. True *(34)*
13. False. If the radiator coolant level is low, the driver/operator should add *approved antifreeze* to the system. *(36)*
14. True *(37)*
15. True *(39)*

Multiple Choice

C.

1. a *(22)*
2. d *(22)*
3. b *(23)*
4. c *(23)*
5. a *(23)*
6. c *(24)*
7. d *(24, 25)*
8. b *(25)*
9. b *(25)*
10. d *(26)*
11. c *(27)*
12. a *(27, 28)*
13. c *(29)*
14. c *(29)*
15. b *(33)*
16. c *(33, 34)*
17. d *(34)*
18. a *(34)*
19. a *(34)*
20. b *(35)*
21. d *(38)*
22. b *(39)*

Identify

D.

1. Commercial driver's license *(25)*
2. Antilock braking system *(34)*
3. Society of Automotive Engineers *(38)*

E. *(40–42)*

1. W
2. D
3. D
4. W
5. D
6. D
7. W
8. D
9. W
10. D

List

F. *(27)*

Answers must include five of the following.

- Visible body damage
- Doors—close tightly, latches work, window glass intact and clean
- Steps, platform, handrails, and ladders
- Saddle fuel tanks
- Tire and wheel—damage, brake dust, fluid leaks, tire inflation, valve stem, tire condition
- Suspension—springs, spring hangers, shackles, U-bolts, shock absorbers

G. *(27–29)*

Answers must include ten of the following.

- Visible body damage
- Under vehicle—front axle, steering, pump piping
- Windshield and wiper blades
- Lights—lenses and bulbs
- Visible damage to audible warning devices
- Pump intake—normally capped, hose firmly attached, intake valve fully closed
- Pump discharge—hose loaded and secured properly, nozzles clean and in place
- Winches—proper operation, damage to associated parts
- Hydraulic rescue tool systems—proper operation, damage, cleanliness
- Fire pump

H. *(29–30)*

Answers must include five of the following.

- Visible body damage
- Wheels and tires
- Splash guards
- Automatic snow chains
- Compartment doors
- Equipment in compartments or on exterior of vehicle
- Hose
- Booster-tank water level
- Reflective striping

I. *(30–31)*

Answers must include five of the following.

- Visible body damage
- Running and warning lights
- Equipment in rear compartment or on outside of apparatus
- Rear compartment door
- Towing attachments
- Hose loads in main hose bed
- Hose bed cover
- Hose bed doors

J. *(32)*

Answers must include five of the following.

- Speedometer/odometer
- Engine speed
- Oil pressure
- Fuel gauge
- Ammeter/voltmeter
- Air pressure
- Coolant temperature
- Vacuum
- Hydraulic pressure

K. *(33)*

Answers must include five of the following.

- Electrical equipment switches
- Turn signal switches
- High-beam headlight switches
- Heating and air-conditioning controls
- Radio controls
- Audible warning device controls
- Computer equipment controls
- Windshield wiper controls
- Windshield defroster controls
- Automatic snow chain control (if applicable)

L. *(35–38)*

Answers must include ten of the following.

- Engine oil level
- Engine air filter
- Emergency shutdown
- Exhaust system
- Radiator coolant level
- Cooling fan
- Windshield washer fluid level
- Battery condition
- Automatic transmission fluid level
- Power steering fluid level
- Brake fluid (hydraulic brake systems)
- Air system
- Belts
- Leaks
- Electrical wiring

Chapter 4 Answers

Matching

A.

1. f *(60)*
2. h *(60)*
3. k *(60–62)*
4. b *(62)*
5. d *(64, 65)*
6. m *(65)*
7. l *(65)*
8. a *(65)*
9. j *(69)*
10. n *(70)*
11. g *(71)*
12. i *(72)*
13. c *(72)*

True/False

B.

1. True *(46)*
2. True *(47)*
3. False. Homebuilt vehicles are *more likely to have design problems* than those built by a fire apparatus manufacturer. *(48)*
4. False. The driver/operator should start the vehicle *as soon as possible* so that it is warmed up when the rest of the crew is ready to respond. *(50)*
5. True *(50)*
6. True *(56)*
7. False. Jump seats equipped with safety bars or gates and seat belts are *not acceptable substitutes for safety procedures that require firefighters to ride in safe, enclosed positions wearing their seat belts. (58)*
8. True *(60)*
9. False. The extra weight carried on most fire apparatus *can contribute to skidding and rollover* due to excessive weight transfer. *(63)*
10. False. The most common causes of skids involve *driver error. (63)*
11. True *(64)*
12. True *(65)*
13. False. A siren operating on an emergency vehicle moving at 60 mph *(97 km/h)* is audible only 12 feet *(3.7 m)* or less in front of the vehicle. *(67)*
14. True *(69)*
15. True *(70)*

Multiple Choice

C.

1. b *(45)*
2. b *(45, 46)*
3. d *(47)*
4. c *(48)*
5. d *(48)*
6. d *(48)*
7. a *(49)*
8. b *(49)*
9. c *(50)*
10. b *(54)*
11. d *(54, 55)*
12. c *(54)*
13. d *(54)*
14. b *(55)*
15. a *(55)*
16. d *(56)*
17. a *(56)*
18. d *(56)*
19. b *(56)*
20. c *(56, 57)*
21. c *(57)*
22. d *(58)*
23. a *(58, 59)*
24. b *(60)*
25. c *(60)*
26. a *(62)*
27. b *(62)*
28. c *(62)*
29. d *(63)*
30. a *(64)*
31. d *(65, 66)*
32. b *(66)*
33. d *(66)*
34. d *(66, 67)*
35. c *(68)*
36. a *(71)*
37. d *(71)*
38. b *(72)*
39. d *(72)*
40. c *(73)*

Identify

D.

1. Privately owned vehicle *(46)*
2. Emergency medical services *(46)*
3. Antilock braking system *(64)*
4. Automatic traction control *(65)*

E.

2, 3, 6, 7 *(74)*

Chapter 5 Answers

Matching

A.

1. h *(84)*
2. b *(88)*
3. g *(89)*
4. d *(89)*
5. a *(95)*
6. f *(96)*
7. j *(96)*
8. e *(97)*
9. i *(97)*

True/False

B.

1. True *(76)*
2. True *(77)*
3. False. Whenever possible the driver/operator should position the apparatus *upwind* of the incident. *(78)*
4. True *(79)*
5. False. Most of the time a fire hydrant is *located very close* to the fire department connection. *(82)*
6. False. Pumpers supporting sprinklers or standpipes *must sometimes give priority to other apparatus,* especially aerial apparatus. *(82)*
7. False. Drafting operations are common in rural areas but *can just as easily be performed* at urban incidents. *(83)*
8. True *(84)*
9. False. IFSTA recommends *never* using hard suction hose to connect a pumper to a fire hydrant, because *it is not designed to be used under positive pressure. (84)*
10. True *(86)*
11. True *(87, 88)*
12. True *(88)*
13. False. Because the apparatus is usually being operated close to the fire, *the driver/operator's vision is often severely obscured* by smoke. *(90)*
14. True *(94)*
15. False. In level II staging, the incident commander summons additional assistance *through the staging officer. (97, 98)*
16. True *(100)*
17. False. It is considered safe for workers to *briefly enter the warm zone without special protective clothing,* unless assigned a task requiring increased protection. *(101)*
18. False. The command post, staging area, and triage treatment area are located within the *cold zone* at a hazardous material incident. *(101)*

Multiple Choice

C.

1. a *(77)*
2. a *(77)*
3. d *(77, 78)*
4. c *(78, 79)*
5. b *(80)*
6. b *(80)*
7. a *(80)*
8. c *(81)*
9. c *(82)*
10. b *(83)*
11. c *(84)*
12. d *(85)*
13. a *(85)*
14. c *(86)*
15. d *(86)*
16. b *(86)*
17. b *(87)*
18. c *(88)*
19. a *(89)*
20. b *(89)*
21. c *(90)*
22. d *(90)*
23. d *(91)*
24. c *(92, 93)*
25. a *(93)*
26. c *(94)*
27. b *(95)*
28. d *(96)*
29. b *(96)*
30. b *(96)*
31. a *(97)*
32. d *(98, 99)*
33. a *(101)*
34. c *(101)*
35. c *(101)*
36. b *(102)*

Chapter 6 Answers

Matching

A.

1. e *(104)*
2. h *(104)*
3. a *(104)*
4. i *(104)*
5. f *(105)*
6. d *(105)*
7. b *(108)*
8. g *(108)*

B.

1. c *(110)*
2. d *(110)*
3. b *(110)*
4. f *(111)*
5. a *(110)*

C.

1. d *(108)*
2. i *(108)*
3. n *(111)*
4. k *(111)*
5. g *(111)*
6. l *(112)*
7. a *(113)*
8. j *(113)*
9. o *(113)*
10. c *(113)*
11. h *(113)*
12. m *(114)*
13. f *(116)*
14. b *(117)*

D.

1. d *(120)*
2. g *(120)*
3. b *(120)*
4. f *(120)*
5. a *(120)*
6. e *(120)*

True/False

E.

1. True *(104)*
2. False. Water cannot be seen in its vaporous form. It only becomes visible *as it begins to condense. (104)*
3. True *(104)*
4. True *(105)*
5. False. The speed with which water absorbs heat *increases* in proportion to the water surface exposed to the heat. *(106)*
6. False. If a material is water soluble, such as alcohol, water's smothering action is *not likely to be effective. (108)*
7. True *(111)*
8. False. A truly static pressure *is seldom found* in municipal water systems; nonetheless, the pressure in a water system before water flows from a hydrant is *considered* static pressure. *(112, 113)*
9. True *(113)*
10. False. Pumpers must work *harder* at high altitudes because *the less dense atmosphere reduces the pumper's effective lift* when drafting. *(113)*
11. True *(114)*
12. True *(115)*
13. False. The tendency of *different hose brands to expand differently under pressure* affects the velocity and, therefore, affects friction loss. *(115)*
14. True *(116)*
15. True *(117)*
16. False. *Water department officials* should be considered the experts in water supply problems. *(117)*
17. False. Most communities us a *combination* of the direct pumping and gravity systems. *(118)*
18. False. Fire department personnel should note the condition and accessibility of valves during fire hydrant inspections and then *notify the water department if any valves need attention. (121)*
19. False. An OS&Y valve's stem is *outside* the yoke when the valve is *open* and *inside* the yoke when the valve is *closed. (121, 122)*
20. True *(122)*
21. True *(122)*
22. False. Some materials offer *considerably less resistance* to water flow than others. *(123)*
23. True *(123)*
24. True *(124)*
25. True *(124)*
26. True *(124)*

6

Multiple Choice

F.

1. c *(104)*
2. c *(104)*
3. b *(104)*
4. a *(105, 106)*
5. d *(107)*
6. d *(107)*
7. c *(108)*
8. b *(108)*
9. d *(108)*
10. d *(108)*
11. b *(109)*
12. a *(111)*
13. c *(112)*
14. a *(112)*
15. b *(113)*
16. a *(114)*
17. d *(114)*
18. c *(114)*
19. b *(114)*
20. a *(114)*
21. d *(114)*
22. c *(115)*
23. d *(116)*
24. c (116)
25. b *(117)*
26. d *(117)*
27. a *(118)*
28. b *(118)*
29. b *(118, 119)*
30. c *(119)*
31. d *(120)*
32. b *(120)*
33. c *(120)*
34. c *(120)*
35. c *(121)*
36. a *(121)*
37. c *(122)*
38. d *(122)*
39. b *(122)*
40. b *(122)*
41. d *(123)*
42. d *(123)*
43. c *(123)*
44. a *(123)*
45. c *(123)*
46. b *(123)*

Identify

G.

1. Pounds per square inch gauge *(111)*
2. Pounds per square inch absolute *(111)*
3. Post indicator valve *(121)*
4. Outside screw and yoke *(121)*
5. Average daily consumption *(123)*
6. Maximum daily consumption *(123)*
7. Peak hourly consumption *(123)*

Chapter 7 Answers

Matching

A.

1. d *(126)*
2. k *(126)*
3. g *(128)*
4. n *(128)*
5. r *(128)*
6. a *(128)*
7. q *(129)*
8. j *(129)*
9. f *(130)*
10. b *(131)*
11. l *(131)*
12. c *(131)*
13. i *(132)*
14. o *(132)*
15. m *(132)*
16. h *(135)*
17. p *(136)*

True/False

B.

1. True *(126)*
2. True *(126)*
3. False. A *wide-angle deflection* produces a *wide-angle* fog and a *narrow-angle deflection* produces a *narrow-angle fog. (127)*
4. False. Impinging stream nozzles usually produce a *wide-angle* fog pattern, but a *narrow pattern is possible. (128)*
5. True *(129)*
6. False. If the driver/operator is unable to supply the proper pressure, actual flow will differ from that indicated at the nozzle. *(129)*
7. True *(130)*
8. True *(130)*
9. False. High-pressure fog nozzles *are best used for* fighting wildland fires. *(131)*
10. True *(131)*
11. True *(132)*
12. False. Detachable ladder pipe streams are *limited to vertical movements.* Horizontal movement of the nozzle would place *dangerous stress* on the aerial ladder. *(133)*
13. True *(134)*
14. True *(136)*
15. True *(136)*

Multiple Choice

C.

1. d *(126, 127)*
2. c *(127)*
3. a *(127)*
4. d *(127)*
5. b *(127)*
6. c *(127)*
7. a *(128)*
8. c *(129)*
9. c *(129)*
10. b *(129)*
11. d *(130)*
12. b *(130)*
13. a *(130)*
14. c *(130)*
15. b *(131)*
16. d *(131)*
17. c *(131)*
18. c *(131)*
19. c *(131)*
20. a *(132)*
21. d *(132)*
22. b *(132)*
23. a *(133)*
24. d *(133, 134)*
25. c *(134)*
26. b *(134)*
27. d *(135)*
28. a *(135)*
29. d *(136)*
30. d *(136)*
31. b *(136)*
32. d *(136)*
33. c *(137)*
34. a *(137)*
35. c *(137)*

Chapter 8 Answers

True/False

A.

1. True *(140; 166)*
2. False. The *most accurate method* determines friction loss *through true field tests. (142; 168)*
3. False. The figures derived by that formula provide *unrealistically high friction loss figures* for today's hose. *(142; 168)*
4. True *(142; 168)*
5. False. Conduct tests on hose that is *in service,* not on hose that has never been put into service (unless it is about to be placed in service). *(143; 169)*
6. False. Test only *one type of hose at a time;* different hoses have *varying amounts of friction loss* due to differences in construction, fabrics, rubber liners, couplings, and wear. *(143; 169)*
7. True *(144; 170)*
8. True *(147; 173)*
9. False. Total pressure loss is the combination of friction loss and *elevation pressure. (149; 175)*
10. True *(155; 181)*
11. True *(161, 162; 187, 188)*

Multiple Choice

B.

1. b *(140; 166)*
2. c *(141; 167)*
3. c *(141; 167)*
4. d *(141; 167)*
5. a *(141; 167)*
6. a *(144; 170)*
7. d *(144; 170)*
8. b *(144; 170)*
9. c *(147; 173)*
10. d *(147; 173)*
11. b *(148; 174)*
12. a *(149; 175)*
13. a *(155, 156; 181, 182)*
14. d *(159; 185)*
15. b *(159; 185;)*
16. c *(164; 190)*

Calculate (US)

C.

1. 15 psi *(142, 143)*
2. 39.5 psi *(148)*
3. 30 psi *(148)*
4. 41.6 psi *(150)*
5. 117.5 psi *(150)*
6. 75.3 psi in each hoseline *(151)*
7. 34 psi *(152)*
8. 130.4 psi *(152–154)*
9. 122.2 psi *(154, 155)*
10. 71.5 psi *(156–158)*
11. 75 psi *(159–160)*
12. 261.7 psi *(161, 162)*
13. 176.3 psi *(163)*

Calculate (Metric)

C.

1. 307 kPa *(168, 169)*
2. 390 m *(174)*
3. 210 kPa *(174)*
4. 432 kPa *(176)*
5. 1 080 kPa *(176)*
6. 233 kPa *(177)*
7. 357 kPa *(178)*
8. 1 088 kPa *(178–180)*
9. 408 kPa *(180, 181)*
10. 430 kPa *(182–184)*
11. 559.6 kPa *(185–186)*
12. 2 179 kPa *(187, 188)*
13. 1 347 kPa *(189)*

Identify

D.

1. Total pressure loss *(149; 175)*
2. Pump discharge pressure *(161; 187)*

Chapter 9 Answers

Matching

A.

1. b *(192)*
2. d *(194)*
3. a *(195)*

True/False

B.

1. True *(192)*
2. True *(192)*
3. True *(193)*
4. True *(194)*
5. False. Programmable calculators that can be preprogrammed for fireground calculations are *available and inexpensive. (195)*
6. False. The amount of friction loss calculated using the condense "Q" formula for 3-inch hose will be *20 percent greater* than if the same situation is calculated using FL = CQ^2L. *(199)*
7. True *(200)*

Multiple Choice

C.

1. c *(193)*
2. a *(193)*
3. b *(193)*
4. b *(193)*
5. d *(193)*
6. d *(194)*
7. c *(194)*
8. b *(194)*
9. a *(194)*
10. c *(196)*
11. d *(196)*
12. b *(197)*
13. b *(197)*
14. c *(197)*
15. c *(197, 198)*
16. c *(199)*
17. d *(199)*
18. a *(199)*
19. b *(199)*
20. a *(200)*
21. d *(200)*
22. c *(200)*
23. c *(200)*
24. b *(201)*
25. d *(201)*

Chapter 10 Answers

Matching

A.

1. h *(204)*
2. p *(204)*
3. j *(205)*
4. b *(206)*
5. s *(207)*
6. p *(208)*
7. i *(208)*
8. a *(208)*
9. o *(210)*
10. l *(210)*
11. f *(216)*
12. t *(216, 217)*
13. d *(224)*
14. m *(225)*
15. r *(227)*
16. g *(230)*
17. n *(236)*
18. k *(237)*
19. e *(239)*

True/False

B.

1. False. Most modern municipal water systems *are unable to* maintain adequate pressure in the hydrant system for effective fire fighting. *(204)*
2. True *(204)*
3. True *(207)*
4. False. Piston pumps are *very susceptible* to wear. *(206)*
5. True *(207)*
6. True *(207)*
7. False. In a centrifugal pump, the larger the impeller's eye the *greater* the flow capacity. *(209)*
8. False. Centrifugal pumps *cannot pump air and are not self-priming. (209)*
9. False. It is better to be in *parallel* than in series because the series mode may make it impossible to supply the necessary volume of water to the attack lines. *(213)*
10. False. These pressures can present *extreme danger to personnel and equipment. (213)*
11. True *(213, 214)*
12. True *(213)*
13. False. If the impeller's hub is worn down, it is possible to *install smaller wear rings* to compensate for the smaller size and maintain the proper clearance. *(215)*
14. True *(215)*
15. True *(215))*
16. True *(218)*
17. False. A lock to prevent the road transmission from being engaged while a front-mounted pump is operating is *a necessity on either a manual or an automatic transmission. (219)*
18. True *(219)*
19. True *(219)*
20. True *(221)*
21. True *(222)*
22. False. According to NFPA 1901, enough 2½-inch *(65 mm)* or larger discharge outlets must be provided to flow *the rated capacity* of the fire pump. *(223)*
23. False. Discharges to which 1½, 1¾, and 2-inch *(38 mm, 45 mm, and 50 mm)* handlines are attached must be supplied by at least *2-inch (50 mm)* piping. *(224)*
24. False. The tank fill line can be used to circulate water through the pump to prevent overheating when no lines are flowing, but a circulator valve *is more effective. (225)*
25. True *(226)*
26. False. Simply opening the drain valve *does not* allow all water to drain from booster hoselines. It may be necessary to manually unwind and drain booster hoselines or blow them out with pressurized air during freezing weather. *(228)*
27. True *(228)*
28. True *(230)*
29. False. Electronic governors are *so accurate and so quick to respond* that they virtually eliminate the need for a pump relief valve. *(233)*
30. True *(233)*
31. False. NFPA 1901 *requires* a primer control on the pump operator's panel. *(234)*
32. True *(236)*
33. False. The pump overheat indicator may provide an *audible or visual warning* when the pump overheats. *(237)*
34. False. The pumping engine oil pressure indicator measures *only oil pressure*; it *is not* a measure of the oil level in the crankcase. *(236, 237)*

Multiple Choice

C.

1. c *(205)*
2. d *(205)*
3. a *(205)*
4. d *(205)*
5. a *(205)*
6. c *(205)*
7. b *(206)*
8. b *(206)*
9. c *(206)*
10. a *(207, 208)*
11. c *(208)*
12. d *(208)*
13. b *(208, 209)*
14. d *(209)*
15. a *(209)*
16. c *(210)*
17. c *(210)*
18. a *(210)*
19. c *(212)*
20. b *(213)*
21. c *(213–215)*
22. d *(215)*
23. d *(215)*
24. a *(216)*
25. a *(216)*
26. c *(216)*
27. b *(216)*
28. a *(216)*
29. a *(216)*
30. c *(217)*
31. b *(217)*
32. d *(217)*
33. b *(218)*
34. d *(218)*
35. c *(218)*
36. a *(218)*
37. c *(219)*
38. a *(219, 220)*
39. b *(220, 221)*
40. a *(221)*
41. d *(221)*
42. b *(221)*
43. b *(222)*
44. c *(223)*
45. b *(223)*
46. a *(224)*
47. d *(224)*
48. c *(225)*
49. c *(225)*
50. d *(225)*
51. b *(227)*
52. b *(227)*
53. a *(227, 228)*
54. d *(228)*
55. d *(228)*
56. d *(228)*
57. c *(230)*
58. b *(230)*
59. a *(232)*
60. b *(233)*
61. c *(233)*
62. c *(234)*
63. a *(234)*
64. d *(234)*
65. b *(235)*
66. d *(235)*
67. c *(235)*
68. a *(236)*
69. d *(236)*
70. b *(236)*
71. d *(237, 238)*
72. a *(239)*
73. c *(239)*
74. c *(240)*

Chapter 11 Answers

Matching

A.

1. c *(242)*
2. j *(248)*
3. k *(248)*
4. b *(251)*
5. g *(251)*
6. i *(253)*
7. h *(257)*
8. d *(260)*
9. f *(261)*
10. a *(270)*

True/False

B.

1. True *(243)*
2. True *(243)*
3. False. Most driver/operators operate *solely from the onboard water tank* at the vast majority of incidents. *(247)*
4. True *(247)*
5. False. If the nozzles are open when the driver/operator puts the pump into operation, he can throttle up the pump to bring pressure to the proper level with a minimum of guesswork. *(248)*
6. True *(250)*
7. True *(250)*
8. False. The hydrant closes to the fire *may not supply the needed amount of water* or may be *too close* to the fire, endangering equipment and personnel. *(250)*
9. False. The three-hundred-foot maximum length for a forward lay supply line working directly from hydrant pressure does not apply when using *4-inch (100 mm)* or larger supply hose. *(253)*
10. True *(253)*
11. False. A four-way hydrant valve *is not necessary* when reverse laying a supply hose. One can be used if it is expected that the pumper will later disconnect from the supply hose and leave the hose connected to the hydrant. *(255)*
12. False. Water *may be discharged* from a booster line to keep the pump from overheating. *(257)*
13. True *(258)*
14. True *(258, 259)*
15. True *(259)*
16. False. Increasing the size of the intake hose can *increase* the capacity of a pump. *(260)*
17. True *(261, 262)*
18. False. A small pond's rate of replenishment *may make up for lack of volume,* and a small stream *can provide a good supply* if the water is moving rapidly. *(262)*
19. True *(263)*
20. True *(264)*
21. False. The pump should not be engaged until *all connections are made and it is ready to put into operation. (265)*
22. False. If space around the actual drafting spot is limited, the driver/operator should *park the apparatus short of the actual drafting site, make the connections, and then ease the apparatus into the final drafting spot* while other firefighters carry the hose and strainer. *(265)*
23. True *(266)*
24. False. Before the driver operator begins priming, all drains and valves should be *closed* and all unused intake and discharge opening should be capped *to make the pump as airtight as possible. (266)*
25. True *(266)*
26. True *(268)*
27. True *(268)*
28. False. Operating from draft is the most demanding type of pump operation, *both from the standpoint of the apparatus and the operator. (268)*
29. False. This *may not be immediately apparent,* as the vacuum is removed when the hose is disconnected and *the liner may return to its proper position. (269)*
30. False. The water supply for a sprinkler system is designed to supply *only a fraction* of the sprinklers installed on the system. *(269)*
31. True *(269, 270)*
32. False. In most cases, it is desirable to *confirm the presence of a fire* before pumping into the system. *(270)*
33. True *(272)*

Multiple Choice

C.

1. b *(242)*
2. c *(243)*
3. c *(244)*
4. b *(246)*
5. d *(245, 246)*
6. a *(246)*
7. b *(246)*
8. c *(247)*
9. a *(248)*
10. d *(249)*
11. d *(250)*
12. c *(250)*
13. a *(250)*
14. b *(250)*
15. a *(250)*
16. d *(251)*
17. c *(252, 253)*
18. d *(253)*
19. a *(253, 254)*
20. b *(255, 256)*
21. a *(256, 257)*
22. d *(257)*
23. b *(258)*
24. d *(258)*
25. c *(258)*
26. c *(259)*
27. d *(259)*
28. b *(259)*
29. c *(260)*
30. d *(260)*
31. a *(260)*
32. a *(260)*
33. d *(260, 261)*
34. b *(261)*
35. b *(261)*
36. c *(261)*
37. d *(261)*
38. a *(262)*
39. c *(262)*
40. d *(262)*
41. c *(263)*
42. c *(263)*
43. b *(264)*
44. d *(264)*
45. a *(264)*
46. b *(264)*
47. c *(265)*
48. d *(265)*
49. b *(265)*
50. b *(266)*
51. c *(266)*
52. a *(266)*
53. a *(267)*
54. d *(267)*
55. d *(267)*
56. c *(267)*
57. b *(267)*
58. c *(268)*
59. a *(268)*
60. b *(268)*
61. d *(269)*
62. b *(270)*
63. b *(270)*
64. b *(270)*
65. c *(270, 271)*
66. d *(271)*
67. a *(272)*
68. c *(272)*
69. b *(272)*
70. a *(272)*

Chapter 12 Answers

Matching

A.

1. f *(296)*
2. c *(296)*
3. h *(296)*
4. b *(297)*
5. i *(298)*
6. a *(299, 300)*
7. d *(307)*
8. g *(307)*

True/False

B.

1. False. *All driver/operators* must be familiar with static water supply sources because *natural disasters may render the municipal water system inoperable. (296)*
2. True *(296)*
3. False. The angle of the intake hose *does not affect* the height of possible lift. *(297)*
4. True *(297)*
5. True *(299)*
6. False. The two primary factors when evaluating a natural static water supply source are adequacy and *accessibility. (301)*
7. False. Fire departments *can avoid accessibility problems by using* public boat launching facilities for drafting sites. *(303)*
8. False. When an inaccessible water supply is *the only source*, it may be necessary to position portable pumps at that source. *(303)*
9. True *(304)*
10. True *(307)*
11. True *(307)*
12. False. If the pool is *inaccessible to the apparatus,* floating or stationary portable pumps should be used to pump the water to the apparatus. *(308)*
13. True *(308)*
14. True *(309)*
15. False. Pumps should be flushed with clear water to remove excess chlorine after pumping from a swimming pool. *(309)*

Multiple Choice

C.

1. c *(297)*
2. b *(297)*
3. b *(297)*
4. a *(297)*
5. d *(297)*
6. c *(298)*
7. a *(298)*
8. d *(298)*
9. b *(299)*
10. b *(299)*
11. c *(299)*
12. c *(300)*
13. d *(300, 301)*
14. a *(302)*
15. c *(302)*
16. b *(302)*
17. d *(304)*
18. d *(305, 306)*
19. c *(306)*
20. d *(307)*
21. a *(308)*
22. a *(308, 309)*
23. c *(309)*
24. b *(309*

Chapter 13 Answers

Matching

A.

1. f *(311)*
2. c *(312)*
3. g *(312)*
4. e *(312)*
5. b *(313)*
6. d *(317)*

True/False

B.

1. True *(312)*
2. False. Relay pumping operations can use *either medium or large* diameter hose. *(312)*
3. True *(313)*
4. True *(316)*
5. False. The two basic designs for relay pumping operations are *maximum distance* and *constant pressure. (317)*
6. True *(319)*
7. False. When one of the attack lines is shut down, an alert operator can *open the dump line* to prevent a dangerous pressure buildup in the relay. *(322)*
8. False. Automatic pressure control devices are *essential* in a relay operation *because of the cumulative effect of pressure changes. (322)*
9. True *(323)*
10. True *(323)*
11. Relay operations should be shut down in order *from the fire scene to the supply source. (323)*

Multiple Choice

C.

1. d *(311, 312)*
2. a *(313)*
3. a *(314)*
4. d *(314)*
5. b *(315)*
6. b *(316)*
7. a *(316)*
8. d *(316)*
9. a *(317)*
10. d *(317)*
11. b *(318)*
12. b *(317, 318)*
13. d *(317, 318)*
14. c *(319–321)*
15. b *(320)*
16. d *(321, 322)*
17. b *(322)*
18. a *(322)*
19. c *(322)*
20. d *(323)*

Identify

D.

1. Medium-diameter hose *(312)*
2. Large-diameter hose *(312)*

Chapter 14 Answers

Matching

A.

1. f *(329)*
2. j *(330)*
3. h *(330)*
4. b *(330)*
5. l *(330)*
6. i *(331)*
7. d *(332)*
8. a *(334)*
9. m *(335)*
10. k *(344)*
11. c *(345)*
12. e *(346)*

True/False

B.

1. False. *Depending on the dump site's proximity* to the emergency scene, the dump site pumper *can also be* the attack pumper. *(330)*
2. False. NFPA 1901 requires that tenders be designed to be filled at a rate of at least *1,000 gpm (4 000 L/min). (331)*
3. True *(331)*
4. True *(332)*
5. False. Trucks that *may be converted* into tenders include petroleum tankers, milk trucks, vacuum trucks, and surplus military vehicles. *(332)*
6. True *(333)*
7. True *(334)*
8. False. Jet-assisted dumps are *more* expensive and *more difficult* to operate than large gravity dumps. *(335)*
9. True *(335, 336)*
10. False. It is preferable that these decisions be made in *pre-incident planning. (336)*
11. False. The front and center of the incident *may not always be the best location* for the dump site. *(336)*
12. True *(337)*
13. False. A *circular* route is usually the most desirable. *(338)*
14. True *(340)*
15. True *(340)*
16. True *(341)*
17. False. Filling tenders by using overhead pipes and a portable manifold *may take some jockeying and waste time. (343)*
18. True *(343)*
19. False. Filling two tenders at a time is generally *not recommended* unless the fill site pumper is connected to a high-flow hydrant. *(344)*
20. True *(344)*
21. True *(347)*
22. False. Baffles must allow water movement at the bottom of the tank and *air movement at the top to provide adequate venting. (347)*
23. False. The time spent engaging the fire pump and making connections necessary for pumping *generally negates any time advantage. (348)*
24. True *(349)*
25. True *(349)*
26. False. If *four or more tanks* are to be connected using jet siphons, *a second pumper should draft from the tanks* solely to supply some or all of the jet siphon lines. *(350)*
27. True *(352)*
28. True *(353)*

Multiple Choice

C.

1. d *(329, 330)*
2. b *(330)*
3. b *(331)*
4. d *(331)*
5. c *(331)*
6. c *(331, 332)*
7. d *(332)*
8. a *(332)*
9. b *(332)*
10. d *(333)*
11. c *(333)*
12. b *(333)*
13. b *(334)*
14. b *(334)*
15. c *(334)*
16. c *(337)*
17. b *(338)*
18. a *(339, 340)*
19. c *(340)*
20. a *(341)*
21. a *(341)*
22. a *(341)*
23. c *(342)*
24. d *(342)*
25. b *(342)*
26. c *(343)*
27. c *(343)*
28. b *(343)*
29. a *(343)*
30. c *(344)*
31. d *(345)*
32. c *(345)*
33. b *(346)*
34. d *(346)*
35. b *(346)*
36. c *(347)*
37. a *(348)*
38. a *(348)*
39. c *(348)*
40. d *(348, 349)*
41. b *(349)*
42. a *(349)*
43. d *(349)*
44. a *(349, 350)*
45. c *(352)*
46. d *(353)*
47. b *(353)*
48. d *(353)*
49. a *(353, 354)*
50. b *(353, 354)*

Chapter 15 Answers

Matching

A.

1. f *(356)*
2. k *(356)*
3. c *(356)*
4. l *(356)*
5. g *(356)*
6. n *(356)*
7. j *(356)*
8. o *(357)*
9. m *(357)*
10. h *(357)*
11. a *(362)*
12. i *(363)*
13. e *(366)*
14. b *(366)*

True/False

B.

1. True *(356)*
2. True *(356)*
3. False. Some *early* alcohol-resistant concentrations *were* proportioned as high as ten parts of water, *but they are no longer in use. (357)*
4. True *(357)*
5. True *(361)*
6. False. A *short drain* time means *rapid* wetting. A *long drain time* means that the foam *holds water and provides an insulating layer over an extended period* before the water is released. *(362)*
7. True *(362)*
8. False. The application rate for protein and fluoroprotein foams on nondiked hydrocarbon fuel spill fires using portable extinguishing equipment is *0.16 gpm/ft² (6.5 [L/min]/m²)* for 15 minutes. *(363)*
9. False. Unignited spills *do not* require the same application rates as unignited spills because *radiant heat, open flame, and thermal drafts, do not attack the finished foam as they would under fire conditions. (364)*
10. False. Concentrates designed to be used at 3% on hydrocarbon fuels and at 6% on polar solvents are commonly referred to as *three-by-six* concentrates. *(365)*
11. True *(365)*
12. True *(367)*
13. False. Because Class A foam concentrations are normally used at very low concentrations. Installed in-line eductors *are not effective* for proportioning Class A foams. *(369)*
14. True *(370)*
15. True *(371)*
16. False. CAFS proportioners typically *do not use foam eductors* because they are not designed to work at either the 0.1 to 1.0 percent eduction rates or the variable flow rates that Class A foams require. *(374)*
17. True *(374)*
18. False. A 1½-inch *(38-mm)* hoseline may be equipped with a nozzle tip up to 1¼-inches *(29-mm)* in diameter. *(375)*
19. True *(375, 376)*
20. True *(376)*
21. False. The roll-on method is used *only on a pool of liquid fuel* (either ignited or unignited) *on the open ground. (378)*
22. True *(378)*

Multiple Choice

C.

1. b *(355)*
2. a *(357)*
3. c *(357, 358)*
4. a *(358)*
5. b *(358)*
6. c *(358)*
7. d *(359)*
8. b *(359)*
9. a *(359, 360)*
10. d *(360)*
11. d *(361)*
12. b *(361)*
13. c *(361)*
14. b *(361)*
15. c *(362)*
16. a *(362)*
17. c *(362)*
18. d *(363)*
19. c *(363)*
20. b *(363)*
21. b *(364)*
22. c *(364)*
23. b *(364)*
24. d *(365)*
25. a *(365)*
26. d *(365)*
27. c *(366)*
28. d *(366)*
29. b *(366)*
30. c *(366)*
31. c *(367)*
32. b *(367)*
33. c *(367)*
34. a *(368)*
35. c *(368)*
36. d *(368)*
37. c *(368)*
38. b *(369)*
39. d *(369, 370)*
40. c *(370)*
41. a *(370)*
42. a *(370, 371)*
43. d *(371)*
44. b *(372)*
45. a *(372, 373)*
46. d *(372, 373)*
47. b *(373)*
48. d *(373)*
49. c *(373, 374)*
50. a *(374)*
51. d *(374)*
52. b *(375)*
53. d *(375)*
54. c *(375)*
55. c *(375)*
56. b *(376)*
57. a *(376)*
58. d *(377)*
59. b *(378)*
60. d *(378)*
61. a *(378)*

15

Identify

D.

1. Compressed air foam system *(361)*
2. Film forming fluoroprotein foam *(365)*
3. Aqueous film forming foam *(365)*
4. Jet ration controller *(368)*
5. Water expansion system *(373)*
6. Water expansion pumping system *(373)*

Chapter 16 Answers

Matching

A.

1. e *(380)*
2. k *(381)*
3. i *(381)*
4. m *(382)*
5. c *(384)*
6. h *(388)*
7. n *(391)*
8. a *(392)*
9. l *(395)*
10. d *(395)*
11. g *(395)*
12. j *(396)*
13. b *(396)*

True/False

B.

1. True *(380)*
2. True *(380)*
3. True *(380)*
4. False. Fire department personnel generally *are not involved* in performing manufacturer's certification tests. These tests are performed by the manufacturer or Underwriters Laboratories personnel. *(380)*
5. False. Road tests should be conducted with the *water and/or foam tanks full* and the *weight of hose and equipment that will be carried on the apparatus accounted for*. *(380)*
6. True *(382)*
7. False. Most departments prefer to draft from *a static source* during pumper service tests because this is a true test of the pump's ability. *(382)*
8. True *(385)*
9. False. Flowmeters *may be used* instead of a pitot gauges; they allow much more flexibility and help to complete tests more quickly. *(387)*
10. True *(389)*
11. True *(393)*
12. False. To test the flowmeter, a hoseline equipped with a solid stream nozzle must be connected to each discharge being tested; *they do not all have to be done at once*. *(393)*
13. False. Tank-to-pump flow tests should be performed *for the higher figure* if one greater than the NFPA minimum was specified when the apparatus was ordered. *(393)*
14. True *(394)*
15. True *(396)*

Multiple Choice

C.

1. b *(380)*
2. b *(380)*
3. a *(380)*
4. d *(380)*
5. b *(380)*
6. c *(381)*
7. b *(381)*
8. c *(381)*
9. d *(381)*
10. a *(381)*
11. c *(382)*
12. a *(382)*
13. c *(382)*
14. b *(382)*
15. d *(382)*
16. c *(383)*
17. a *(383)*
18. c *(383)*
19. d *(384)*
20. b *(384)*
21. b *(384–386)*
22. d *(384–386)*
23. d *(385)*
24. c *(387)*
25. b *(389)*
26. b *(393)*
27. c *(393)*
28. d *(393)*
29. c *(394)*
30. d *(394)*
31. a *(395, 396)*
32. b *(396)*
33. d *(397)*

Distinguish

D. *(385)*

1. N
2. R
3. R
4. N
5. R
6. N
7. N
8. R